THE WHOLISTIC HEALING GUIDE TO CANNABIS

UNDERSTANDING the Endocannabinoid System

ADDRESSING Specific Ailments & Conditions

MAKING Cannabis-Based Remedies

TAMMI SWEET

Storey Publishing

The mission of Storey Publishing is to serve our customers by publishing practical information that encourages personal independence in harmony with the environment.

Edited by Liz Bevilacqua
Art direction and book design by Michaela Jebb
Text production by Jennifer Jepson Smith
Indexed by Samantha Miller

Cover photograph by © Michael D. Wilson

Interior photographs by © Tammi Sweet, except © Andrea Obzerova/Alamy
    Stock Photo: 37 bottom, © blickwinkel/Alamy Stock Photo: 29, and
    © Michael D. Wilson: 35

Cover and interior illustrations by © Sally Caulwell (instagram@sallycaulwell)

This publication is intended to provide educational information on the covered subject. It is not intended to take the place of personalized medical counseling, diagnosis, or treatment from a trained health professional. Please be aware of cannabis-related laws that may apply to you. Please exercise caution when using health remedies of any kind and follow all safety guidelines.

Storey books are available at special discounts when purchased in bulk for premiums and sales promotions as well as for fund-raising or educational use. Special editions or book excerpts can also be created to specification. For details, please call 800-827-8673, or send an email to sales@storey.com.

Storey Publishing
210 MASS MoCA Way
North Adams, MA 01247
storey.com

Printed in the United States by Versa Press
10 9 8 7 6 5 4 3 2 1

Library of Congress Cataloging-in-Publication Data on file

CONTENTS

Gratitude

I begin with my plant ally, cannabis. Thank you for all the ways you have helped me to be a better human. Thank you for your generous gifts that allow me to be here, sharing your magic with others. I hope I have represented you in ways that are helpful and respectful.

Next, I come to the human plant-loving clan I call family, the herbalists. My first visit to this world was at the New England Women's Herbal Conference in the late 1990s. Although I initially felt like a tree among the flowers, with my basketball shorts and jock attitude, I soon knew these were my people. It was here that my understanding of "right relationship" blossomed and was nourished. Elders honored for the accumulated wisdom that comes from living a full life held council and told stories under an old oak while the line of budding herbalists brought their gifts and gratitude. We understand our abundant life is made possible by the life-giving generosity of our ancestors, the green world. We would not be here without their offering of the oxygen we breathe, food, clothing, shelter, and medicine. Rosemary Gladstar, my beloved first herbal teacher, exemplifies right relationship as she walks the world spreading fairy dust over those who come close enough to listen. One of her many gifts is her ability to see potential leaders and teachers and offer them a place at the table. Her request for me to teach at Sage Mountain began my path to this book. Pam Montgomery continued my education on the subtle nature of plant spirits and how we must step into our rightful place as healthy co-creators with the plant world. Again, this book would not be here without her mentoring, exploring, and tracking the places of plant relationship and her generosity of spirit to share what she found. Tom Brown Jr., the master mapmaker of the spirit world, thank you for showing me ways for deep connection with spirit.

It takes a village to write a book. Life carries on while I sit in my "Tam cave" researching, learning, and writing. The garden continues to thrive and be nourished, the medicine continues to be made, and chores on the land continue to get done. Thank you, Suzanne, for taking the lead on nourishing me and our land and Kim and Milo and all the interns who kept all the chores rolling. Thank you to Kim for being my personal article retriever and Endnote manager.

Finally, my beloved companion on this wild ride, my wife, Kris. Thank you for holding all the "beta-chirps," all the details of our very full life so I could go and write. For the counseling sessions as I cried about all the animal experiments I needed to sift through in researching this book, for the kind listening to my ideas, for your insightful editing, and for the fun paths you sent me down in exploring new ideas. I wouldn't be here without your solid, gentle, and wise companionship. Thank you.

WELCOME

I would like to introduce you to the plant cannabis.
My intention is for this book to offer meaningful, reliable information and insight about cannabis for the beginning explorer as well as the experienced practitioner of the healing arts. You may be an herbalist or acupuncturist; a nurse, physician, or massage therapist; or psychotherapist, addiction and recovery counselor, or dispensary worker; or a traveler on the path of being human in a challenging world. In any case, we are all looking for reliable information to help people who are suffering find relief.

Perhaps you've heard that CBD oil can cure almost everything. Or a loved one has been suffering from a long-term illness and you've heard cannabis might help. Perhaps you're a parent whose child has struggled with epilepsy or a person who simply feels stuck in a rut of negative thinking. You might have just discovered this plant or already be in a long-term relationship with cannabis. You may have come to this book in a hurry, eager to research cannabis's properties or to make a medicine. Or you may have wandered in with a mind full of curiosity about what's fact and what's mythology. Whatever brings you here, my first choice is to have you land in a comfortable spot in front of a warm fire or a sunlit window with the just-right fullness of a well-nourished body and time to spare.

Cannabis is a remarkable plant with a capacity to teach, to help alleviate suffering, and in some situations to transform and heal. This powerful and wise ally can influence our health and well-being in body, mind, heart, and spirit.

Cannabis is a friendly plant to people. It is no surprise that humans have been working with her for centuries. In this book, you'll discover how her gifts are uniquely matched and resonant with our physiology to make her such an easy and important ally.

Cannabis is also complex: multicolored and subtle, intelligent and diversely adaptive, gentle to use in one situation and requiring a low dose and mindfulness of cultivar-specific sensitivity in another. While chamomile, for example, is vastly talented as a healing herb, it is very difficult to drink chamomile tea with unwanted consequences. However, cannabis, while safe and friendly, is complex. For this reason, it's important to come to your relationship with her with curiosity and openness, with humility and patience and a willingness to listen — perhaps as you would with someone you want to form a meaningful relationship with.

This is not intended to be a drive-thru experience or a speed-dating session. While I've attempted to write in ways that are user-friendly and accessible across differences of experience and knowledge, this plant requires and deserves a more thoughtful inquiry, and I strongly encourage you to settle in for a journey. While some questions may be answered by a quick visit to the index or a glance at diagrams, many require a fuller exploration about intake methods, dose, context, and yes, intention.

As a scientist and herbalist, and as an ordinary learning, evolving human, I have been astounded by the spectrum of gifts this plant makes available to people. I'll spend time discussing the important benefits of cannabis in diverse physical health situations in the chapter on conditions (page 164). It takes my breath away to discover how many people in diverse situations have been helped by this plant through complementing other interventions or when other interventions have failed or created negative effects of their own.

Other benefits of cannabis are also addressed, such as helping us come into present time, letting go of destinations and goals, nourishing creativity and exploring new ideas, even connection and play, all within a container of safety. At a personal level, the most profound benefits for me have been to my mental, emotional, and spiritual self. The shifts made possible in my consciousness and self-awareness, and the kind and insightful assistance cannabis has offered, have helped me transform limiting personal beliefs and response patterns with remarkable ease. I didn't understand the physiology at the time, but by creating a state of neurochemical safety, the plant supported me to reprogram habits and reflexes that I had learned as an adaptive child but that no longer served me.

There are other ways that cannabis is restorative to what is wise and healthy in us as children that has been forgotten or conditioned away. For example, as a child, I loved the wild world of nature and the experiences of exploration and adventure. My boundaries were set each year by landmarks within the woods. My mom would walk with me out there and say, "This is how far you can go this year." In the beginning, feet felt like miles and time had no clock. I still remember being facedown in the grass looking at bugs. The company of my favorite quiet hemlocks. The deer family walking by in their daily routine undisturbed by my presence. Out where I could be quiet and just listen. Out where I felt at home by the little vernal pool I thought of as a pond.

Eventually, the railroad tracks became my boundaries, and my wilderness area was replaced by a condominium complex. My friends and I tried to protect our sanctuary by removing all the boundary markers. Even then we knew the importance of wilderness and of natural states, of being open and receptive to curiosity and wonder, of exploration and peace of mind and heart.

Being in nature, as a part of nature, heals an underlying ailment of our culture — our disconnection with our true home. It enables us to remember who we are and what our place is in it all. Cannabis is gifted in helping us remember this and experiencing the states of aliveness of our deeper natures by being in present time.

This kind of emotional/spiritual learning is best done by making informed and thoughtful choices about yourself and the plant cultivar, the context and setting for your experience, and by trusting your capacity to learn and unlearn and grow, to become your wisest self. These journeys are deeply personal and are enhanced by a respectful and humble relationship with the plant.

Since we're embarking on a journey, a bit about your guide . . .

I love puzzles and understanding how things work. From this passion I was drawn as a student to Western science for tools and models of exploring. I love the left-brain scientific world, the magic of mitochondria in a living cell, and the benefits of formulating hypotheses, executing reproducible experiments, and evaluating results for peer review. I love critical thinking. In college, these led me to pursue a master's in endocrinology.

While pleased with my path, I concurrently felt an unfulfilled tug, something missing, in my learning and understanding about the world through this worldview and methodology. I wanted to grasp more about the mind and philosophy and the nature of life in another dimension. This led to pursuing a minor in religion, a path that opened a wider lens for exploration and for explanation.

Over the years, I went on to discover indigenous sciences, and again my lens widened for understanding. While there exists a Western bias that undermines recognition of indigenous knowledge as science, deeper reflection makes it clear that direct communication between plants and humans is ancient and time-tested. Perhaps the longest evidence-based longitudinal study ever done is the fact that humans have survived and thrived through a deep and interdependent relationship with plants since our beginnings on Earth. This outcome is likely only made possible as a result of our attuned senses (including what we'd call intuition), developed and designed to let us know and communicate with plants in an adaptive, evolving, side-by-side relationship. Indigenous science has understood this for a very long time.

My training in each of these realms has let me walk in both paradigms and worlds in hope of helping each understand the other. My dream is to join the many kindred others around the globe who are bridging these worlds to foster a bringing together of collective wisdom in ways that may help alleviate suffering, aid healing, and nourish a healthier and more balanced life for humans on Earth. I believe one of the many gifts of cannabis, and her presence at this particular time, is her ability to help bridge the different paradigms of learning and thinking.

I love that as a teacher I am able to share what I've learned in ways people can understand, remember, and apply to their lives. It is my utmost wish that something within this book will help you discover the remarkable nature of cannabis and how to develop a meaningful relationship with her. I hope that something will offer benefit to you, a friend or family member, or a client or a community. A benefit for body, heart, and/or spirit, whether it be a little less suffering, a little more peace, or a little more freedom from the patterns and habits that prevent you from being a better human.

There is so much to appreciate about our human experience with cannabis. But cannabis is also misunderstood and abused by the same social forces that have negatively influenced human relationships with all of nature: greed, taking without reciprocity, impatience, and choosing quick solutions rather than deep listening and care.

We are embarking on a whole new level of journey with this plant, and I believe we each have a role to play in the quality of life that is generated by our actions in medicine making and production, in research and education, in the market, and most important in growing the plant and working with her gifts. You, the reader, are able to choose how you walk this path of learning and medicine, and you can influence the journey in a meaningful way.

I am deeply grateful for the wisdom and healing that this plant has gifted me. In addition to the ways I am able to help others, I have also become a better human.

Thank you for joining me and for your part in the larger story of life.

Laying the Groundwork

As I work with and teach about cannabis, I am standing on a few foundational principles.

Cannabis Is a Master Plant

Cannabis is not a tonic herb. It's not a plant we take medicinally in large quantities except under very specific conditions. She is a low-dose botanical, not because she is poisonous but because she is a master plant. Like other master plants such as peyote, psilocybin (both actually not plants but fungi), or ayahuasca, cannabis has the ability to change the consciousness of humanity. The term "master plant" comes from shamanic traditions where it is recognized that the spirit of the plant or fungus has the ability to be in direct relation-ship with us and to change human consciousness.

We Are in an Ally Relationship

Plants are our elders. They evolved millions of years before humans and learned lessons on how to survive and thrive on this beautiful planet. We have much to learn from them. When we come to the plants with the respect we bring to our elders, we enter into a sacred relationship. When we are chosen and choose a particular plant to work with, we are in an "ally relationship." We are committing to be schooled and to learn from and with our plant.

I will at times refer to the cannabis plant as "she." First, this is a sign of respect and gratitude. Second, the plants we use for making medicine are female.

Cannabis Offers Plant Spirit Medicine

Spirit medicine is the oldest science, more ancient and tried and true than the Western science we now swim in. We use the skills passed down through the generations by medicine people from all walks of life. Skills embedded in our DNA allow us direct communication with our ancestors, the plants. Spirit medicine asks us to step away from the modern world: our computer, the Internet, even our left brain. As neuroanatomist Jill Bolte Taylor says, "We must step to the right of our left hemisphere" and venture out into the vast conscious-ness of our heart, which connects us to everything.

Whole-Plant Medicine Is the Best Medicine

Whole-plant medicine means using the entire female cannabis flower and trichome-filled leaves to create medicine rather than isolating specific parts of the plant or extracting components to create chemical cocktails. By using the full plant, we benefit from all of the phytochemicals working collectively. Western science acknowledges this benefit; it labels it the "entourage effect." Herbalists call it business as usual.

We Must Be in Right Relationship

Working with cannabis, or any plant, as a plant ally is based on relationship, one that develops over time and can't solely be cultivated with book learning or casual participation. The plant will undoubtedly bring her gifts, and you must bring your whole self, willing to be in a co-creative relationship, not a relationship of dominance and "knowing better." You will need to surrender your ideas about control, specifically who has it. Don't worry; we all have control issues of one kind or another, and you will have plenty of opportunities to practice working on yours with cannabis.

If you are not used to thinking in this way, think of it as building up a very small muscle. It takes time and use to become strong.

Right relationship also means we pay attention to how we use this medicine. If you use cannabis in ways that dishonor the relationship, it will follow the path of other abusive relationships, as has happened with tobacco, poppy, sugar cane, and coca leaf. We now have a cultural understanding of what happens when we are not in right relationship, and it is my profound hope we learn from our experience and bring forward health and healing.

Tammi

CHAPTER ONE

HOW WE GOT HERE

Across the United States, individual states are legalizing cannabis and creating dispensaries, and more and more people are seeking cannabis and CBD oil for healing. Cannabis has been used for thousands of years, but it has rapidly come to the forefront in our culture. People are calling for it — demanding it. This master plant has captured our attention. Why has cannabis emerged in mainstream consciousness recently? I believe that she can teach us how to live in ways that we've forgotten. The plant has wisdom to offer about the ways that we are treating the planet and our bodies. I believe cannabis is here to teach us how to heal. Ourselves. Each other. The planet.

The overarching function of the endocannabinoid system (ECS; see page 65) is to maintain balance. When the system is in balance, it sends a signal throughout the body that all is well. As we get to know the plant, we can understand its integral function in the ECS within our bodies. We will understand how this "global protection system" functions, and we can learn how to nurture and restore this important system of safety and well-being.

A *Herstory* of Cannabis Medicine

Cannabis has been used for health and healing for at least 6,000 years (that we have documentation of). The entire plant is a treasure trove of life-giving richness. The seeds are high in vitamins A, C, E, and beta-carotene. They are rich in protein, carbohydrates, minerals, fiber, and have an ideal ratio of omega-6 to omega-3 fatty acids. The stalks can be used for fiber in clothing, rope, fabric, and as a replacement for plastics and building materials. The flowers have been used medicinally in China, Egypt, and India for some 5,000 years. In one archaeological find, a 2,700-year-old Gobi Desert shaman was unearthed with a pound of unseeded sinsemilla cannabis flowers.

Cannabis has been on the planet for approximately 60,000 years figuring out all kinds of survival strategies. Much speculation exists about its place of origin, and it's almost impossible to nail one down because as soon as humans found this cornucopia of help and vitality they started cultivating her, transporting her seeds, and planting her wherever they traveled. Along with the ambiguity of where the plant originated there is the problem of confusing common nomenclature, which does not match up with the botanical taxonomy. Finally, botanists can't quite agree if there are two species of cannabis or three (I'm going with the two camp). So confusion is everywhere! Traditional medicinal uses in China, India, Egypt, and the United States were for analgesia and anti-inflammatory, antispasmodic, anticonvulsant, and sedative properties. Thousands of years of practical experience with cannabis tell us to start with at least these applications.

Mainstream Medical Practice

Western medical practice would have us believe we can separate the body into parts, isolate certain structures, understand their function, and then prescribe how to fix them. This may be a convenient theory, but in my experience it does not ring true when dealing with a complex living being. Nothing in our body works in isolation. We are complex communities of 50 trillion cells working collectively. It is naïve and dangerous to act and believe as if we can be separated out into isolated organs and systems or that we can separate our physical, emotional, psychological, and spiritual selves and only treat the physical.

We cannot separate the plant into components we deem active or functional and those we deem inactive or unnecessary. The science regarding cannabis as medicine has proven this. In fact, every study on effectiveness shows whole-plant extracts to be far superior to partial-plant isolates, on the order of 4 to 330 times more effective depending on the study. Western medicine calls this the "entourage effect"; herbalists consider this an obvious effect of whole-plant medicine.

Western medicine also operates under the foundational principle that everything must be measurable, reproducible, and labeled. It considers working with plants to be "sloppy" or "vague." Herbalists understand that when you are dealing with a plant with thousands of active constituents, you simply cannot measure the effects of what is happening with an individual constituent. What happens inside one human will not be exactly the same in the next. Western medicine wants the dosing to be consistent or standardized. It would like the effects to be consistent and standardizable. The belief in Western medical practice is that isolating a constituent of any kind of medicine enables the practitioner to zero in and measure what is happening in the individual with that one part.

This is a false assumption by Western medical practitioners. A one-size-fits-all prescription does not work the same in every individual. How could it when the measure for a "standard dose" is for a 150-pound man? What if you are a 200-pound pregnant woman? What about *all* the factors that change every day for our bodies internally, even for the 150-pound man? How's the liver that day? What's the overall level of inflammation? Or *goddess* forbid, what's going on emotionally or spiritually that would have an impact on one's physiology?

But naturally occurring constituents cannot be patented. Isolated constituents can. If you add a hydrogen atom, *voilà!* It's yours to patent and own — and profit from.

The Myth of Standardization

I've thought a lot about the obsession with isolated extracts and standardization both in the context of pharmaceuticals and cannabis medicine. While I understand the need and want for reproducible results and the safety of known quantities for dosing powerful pharmaceuticals, I can't help but consider another seldom-discussed reason: a false sense of control.

Herbalists don't work in standardized extracts. Herbal medicine as an art and a craft requires an additional kind of knowing, a different

CRIMINALIZATION

Up until 1942, cannabis was used in the U.S. pharmacopoeia by the medical establishment. In 1970 it was classified as a Schedule I drug in the United States, which by definition designates it as having "no medicinal value." Although 32 states have created their own laws legalizing the use of cannabis for medicine and/or recreation, the federal law still stands.

The time frame leading up to the banning of cannabis use was rife with propaganda and misinformation. One strategy was to link its use to people of color and immigrants to create fearmongering. Another strategy was to refer to cannabis as "marijuana," because that's what Mexicans called it. The word was meant to be shaming and disrespectful. I will not contribute to the demonization of a people or a plant by using a word meant to hurt. We will refer to this miraculous plant as cannabis.

Thank you to the people who, at risk of incarceration or having been incarcerated, have continued to cultivate, perpetuate, and create the varieties we have access to today.

understanding of how the world works. I'm not suggesting herbalists don't use scientific knowing and understanding; it's just not the only way. Isolated extracts may tame cannabis and make her more "manageable" and knowable, but in a false sense. Whole-plant extracts require trust and understanding, a feminine mode of operation to receive what is given, to receive rather than dominate.

It is true that if you break cannabis apart into small, barely recognizable pieces of herself you can "control" her and feel like you can mitigate cause and effect, but you are left with vestiges of what was once a great power. You feel a whisper of what the plant's capable of, traded for a false sense of control. Practitioners want clear, evidence-based information on standard doses for every person seeking assistance. It is my profound hope that one of the many lessons cannabis has to teach us is that healing is not standardized. One dose does not fit all (or one plant cultivar). We need to know the condition of the person in front of us, the whole person. Getting to know them will require more than a five-page medical history chart and an eight-minute office visit. Healing requires relationship, relationship with the person and an intimate relationship with the medicine you are working with — a history of learning the subtleties and nuances of the specific plant. These things cannot be learned from a pamphlet. They can't be learned overnight, in a four-hour course, or from reading a book. The craft and skill of working with plant medicine is a long-term, intimate relationship the practitioner forms with plants themselves, and like any lasting and deep relationship, it takes time.

WHAT ABOUT FULL-SPECTRUM CBD?

Some CBD oil products are isolates. A full-spectrum CBD oil contains complete, unfiltered extract from the whole plant, including a full range of cannabinoids and terpenes.

The Herbalist's Approach

One cold winter, I decided I wanted to learn how to make beer, so I went out and bought Stephen Buhner's book *Sacred and Herbal Healing Beers: The Secrets of Ancient Fermentation.* I read it cover to cover and jumped right in to homebrewing. The thing I appreciated most — in addition to the tons of research Buhner did — was a section at the back that essentially said, *Don't listen to the doubters; it's not that complicated. Go make beer.* I love this can-do attitude; it permeates the herbal world. My herbal elders and teachers — Rosemary Gladstar and Stephen Buhner — champion our ability to bring health and well-being into our own lives while also promoting right relationship with the plant world. Humans have been working with plant medicine since we've been on the planet. It's in our bones to be in direct relationship with the plant world and the healing it can offer. Learn the basics and go start making your own medicine. It is your birthright.

Whole-Plant Medicine

Whole-plant medicine means just that: the whole plant, and all of its constituents, is used for the medicine. Herbalists hope to preserve as many of the different constituents of a plant as possible, not just the constituents we pick and choose or deem important. We would do well to use the entire chemical profile of the cannabis plant rather than deciding that CBD or THC is the *one* constituent to use, or that the chlorophyll or lipids are not important and should be removed. The science shows what herbalists already understand: that whole-plant extracts are more potent. Study after study shows that whole-plant cannabis extracts are anywhere from 4 to 330 times more effective than single-plant extracts or synthetic medicines. Yes, 4 to 330 times more effective. This is true for the entire range of conditions cannabis works with. We could also say isolated extracts are 4 to 330 times weaker than medicine we make at home.

That doesn't mean we use all parts of the cannabis plant in our preparations. We use all of the parts that are medicinal and all the constituents of these parts. That is, the whole flower and sepal leaves containing the trichomes (the little crystal-like hairs on the buds) for most medicines, or the roots, depending on what kind of medicine we are making. Whole-plant medicine for herbalists means we use all of the plant parts we are working with rather than chemically extracting out isolated constituents. It also means we know why we are using each particular part of the plant.

A Wholistic View

The wholistic approach to healing includes looking at the whole equation, including the person, the condition, the dosage, the cultivar, and the grow environment. Every part of the equation is important and contributes to healing. The practitioner must get to know a person in order to treat them effectively. My teaching philosophy is based on the premise that if practitioners understand *how* something works (herbs, diet, exercise), they can draw on their own creativity to come up with healing solutions that perhaps no one has thought of. Formulas can be a helpful starting place, especially if the *how* is addressed, but the collective wisdom of herbalists is far more intelligent than one person and their one formula.

Cultivating Relationship

Relationship with cannabis means you have spent time researching, working with, and perhaps cultivating and journeying with cannabis — however you relate best. The relationship also requires educated trust in the plant based on your experience with her. As an herbalist, relationship with my clients means not only do I know their history and the current conditions they are working with, but I also try to understand the whole of the person in front of me, including their mental, emotional, and spiritual selves, as well as their physical state. Like any relationship, all of these require time — lots of time. What we offer at the beginning of our relationship will look quite different than after we have spent years becoming intimate with both the plant and the people we work with.

How cannabis plants are grown is vitally important. Small-scale growers are a lot like small-scale beekeepers. Do you know a beekeeper? Ask them about their bees and watch their whole demeanor light up. They love their bees. I want my medicine to originate from to gardening! Many of my students have come to herbalism through cannabis. Cannabis was the first plant they worked with or grew and wanted to get to know better. What a beautiful thing it is to open people to the world of healing plants and how to cultivate relationship with them.

the hands of people who love their plants, who love working with cannabis, who sing and pray and create a healthy, sacred environment for the plants to grow in. People who light up when you ask them about their plants.

Our work with cannabis can also help us relate to the Earth in more respectful ways. In the western United States, where growing cannabis has been legal for a number of years, there is an emergence of consciousness around restoring health and vitality to the agricultural land, called regenerative farming. Working *with* the land to grow plants creates more biodiversity and brings health and well-being to the land — as well as to the animals, insects, microbes, and plants growing in our little gardens.

Practicing Humility

Herbalists recognize that we don't know everything. Knowledge of how plant constituents work together is complex, and Western science often doesn't understand how these combinations work. In the whole-plant herbalist paradigm, we don't need to shun or change the mystery of how plant constituents work together (even if we don't know all the ways). We trust the wisdom of the plant and our experience working with plant medicine. Experience and trust are valued characteristics.

The amazing thing about cannabis and our relationship to it is that a cannabinoid system exists within the plant and a cannabinoid system exists within our bodies (we'll explore the *endo*cannabinoid system within our bodies later in the book). This is a profound connection. This phenomenon is the basis for what herbalists do. It is the foundation for healing with plants. The more we understand about the cannabinoid systems in the plant and in ourselves, the greater potential we have for healing ourselves naturally with the gifts of the earth.

Sourcing Cannabis for Medicine

Cannabis is a plant. Source it as you would any medicinal plant you are going to work with to create therapeutic remedies. It's best to use local, organic, and pesticide-free plants. And it's important to know the conditions the plant was grown in.

If you want to grow cannabis yourself, there are many excellent resources out there to teach you how; I've listed some of them in the Resources section on page 249. Whether you're growing cannabis yourself, buying it from a local grower, or picking it up at a dispensary, it is crucial to know how the plants were grown. There are a range of growing options. As with any gardening, you can choose the easiest and least expensive options or you can invest time and energy in organic growing. For the health of the plants, the grower, the people making and consuming cannabis products, and the health of the planet, I encourage organic practices. As a consumer, we can encourage these practices by asking it of our growers and being willing to pay for the care and time it takes to carry them out.

The Agricultural vs. Horticultural Model, or Takers vs. Leavers

Ask your grower about their approach to growing. Cannabis, like any crop, can be grown using a large-scale agricultural model or a smaller-scale horticultural model.

Large-scale "throw-and-grow" farmers plant hemp like industrial corn. They crowd as many plants as possible into a plot and use petroleum-based fertilizers that ultimately deplete the soil and the life around the fields. While this model may have been developed with good intentions — to feed more people and use new tools and products to increase productivity — it lacks perspective on what we will leave for the next generation and how much we can take from the land and soil without depleting it. Every harvest removes biomatter and nutrients from the land, so we must help farmers transition away from this depleting model into a more sustainable and regenerative one. Some growers may feel increased pressure to get their flowers to market first and therefore harvest early. Harvesting too early can create more of a "zippy" effect in the end-product medicine than a calming one.

A more wholistic model believes that we are part of, and interconnected with, the world around us, and we must work to improve our natural environment, starting with the soil. Biodynamic, permaculture, and regenerative growing philosophies all utilize this belief in their growing practices.

Questions for Your Grower, Dispensary, or Medicine Maker

- How large or small is the grow operation?

- Are the plants grown outdoors or indoors?

- Are the plants grown in soil, a sterile growth medium, or hydroponically?

- What nutrients are used? Are they organic, organic-based, or petroleum-based? If a person cannot answer this, ask for the brand names of the nutrients and look them up.

- Does the grower handle disease and pest management organically or nonorganically? Are these practices applied during flowering?

- Does the grow operation use biodynamic practices?

- Why are they in this business? What is their mission or what are their founding principles?

Sun and Soil

In my ideal world, all cannabis would be grown outdoors in a biodynamic garden alongside other medicinal plants that serve to feed pollinating bees, birds, and insects. Given the current status of state and federal laws regulating cannabis, you may be able to grow plants or obtain plants grown in this way in your state — or you may not.

Cannabis grows best in nutrient-rich, well-drained soil in full sun with little competition. Cannabis plants that are grown locally outdoors in the sun in organic soil rank number one. You can add compost, sand, perlite, vermiculite, mycelial inoculant, or biochar. Cannabis likes well-drained soil, so amendments that increase the draining capacity or nutrient value of the soil are important, especially if you have high-clay soil or live in an area that receives a lot of rain. If you are going to use amendments, it's best to be sure they are organic and a renewable resource, and you should consider any effects they may have on your garden's ecosystem.

You can also use grow media that need added nutrients. Among these, I suggest using coir — shredded coconut shells — as a good growing medium; it is more renewable than peat moss. Peat moss is a standard ingredient in most commercial growing mixes, but it is not a renewable resource; it is removed from peat bogs where it has taken thousands of years to accumulate, and it cannot easily or quickly renew itself.

Indoor Growing: Greenhouse and Hydroponic

Indoor growing may be the only option where there are obstacles, such as a short growing season or limited land access. Indoor growing can yield healthy medicinal cannabis if managed properly. The benefit of growing indoors is that you can monitor humidity and there's a much lower risk of mold or fungus than in outdoor soil. Factors to consider for growing indoors include your light source, light schedule, soil and soil amendments, and insect management.

In greenhouse growing, you can regulate the amount of water absorbed into the soil. The rainy and humid summer and fall seasons we experience in the Northeast can be a perfect storm for mold and mildew. While a greenhouse can decrease the amount of water in the soil, it can still be a high-humidity environment (a breeding ground for mold and mildew), so you'll need a dehumidifier in there.

In hydroponic gardening, water is the growing medium. The grower supplies all the nutrients to circulating water, and plants are grown without soil at all. While hydroponic gardening has the potential for the greatest yield of flowers, it is difficult and has little room for error — not an ideal entry into cannabis growing for beginners.

Nutrients

Nutrient use garners cultlike conversation among growers. People are particular and partial to certain methods. It can sometimes feel based on superstition rather than fact. *"Use this product and twirl three times!"* Often a particular method has worked for a grower, so they stick to it no matter what, sometimes with no explanation other than that it works for them. That doesn't necessarily mean it will work for you.

There are all kinds of marketing ploys for special cannabis-growing nutrients, but cannabis is a plant, and if you get to know the plant you can get a perfectly respectable yield that will meet your medicine-making needs without any special nutrients. You can get gigantic, dense, picture-perfect buds through purely organic techniques, without added nutrients, but it may take some time to understand and master. You can build high quality, nutrient-rich soil in which to grow plants with no amendments, but it takes time, energy, and experimentation, as each strain of cannabis has particular needs for optimum growing. The best growers will walk with humility and curiosity and learn from each harvest.

The spectrum for nutrients ranges from certified organic to organic-based nutrients to petroleum-based nutrients with nearly fluorescent coloring. The petroleum-based products are the cheapest, and the organic products are the most expensive.

Growers using any kind of nutrients often "flush" their plants — either watering them without nutrients or adding chemical flushes — during the two weeks just before harvest. One benefit of flushing, or at least stopping nutrient addition, at this point is that the plants produce more terpenes. I've heard stories of people lighting up dried cannabis flowers and seeing a rainbow in the flames. This is not the result of some way-out trippy weed but instead the result of not properly flushing petroleum-based nutrients at the end of the grow cycle. The nutrients build up in the plant and produce rainbow flames.

Pest and Disease Management

Insects such as spider mites and fungus gnats are inevitable at some point, especially when growing indoors. Ask your local grower how they deal with insects. Some growers support the plant's natural immune system so it can fight off pests itself; others add either organic or chemical pesticides. You should avoid spraying pesticides on cannabis while it is flowering; if you do, the pesticides remain on the flowers and will be in whatever medicine you make from them. That's not recommended even if the pesticides are organic.

Fungus growth also needs to be managed, especially in outdoor grows where humidity cannot be controlled. Here, too, there are organic and nonorganic practices for managing fungi.

Large indoor grow operations may have fifty thousand plants and can't afford a spider mite or fungal infestation of the crop. They will often spray prophylactically to avoid infestation. These flowers can end up at dispensaries or in a bag from your local supplier. Currently, there are no federal industry standards for pesticide use in the legal cannabis markets. Some states are testing for pesticide levels in dispensary flowers, but no legal limit has been determined. In a recent study, 85 percent of all cannabis flowers from dispensaries in Colorado and 20 percent of flowers in California tested positive for pesticides. Truthfully, we don't know what amount of pesticides are safe to consume. As medicine makers or medicine takers, it is important to consider the implications of pesticide levels, especially in products like concentrated resin extracts, where levels can be considerably higher. Don't assume that growers test their products for pesticide levels. It is smart to ask to see results of testing or test yourself.

Tending and Yield

Before the introduction of pesticides and herbicides to food crops, the average net loss of crops was about 25 percent. I think of this not so much as a loss but as a payment to the surrounding life — the soil, mycelium, bacteria, insects, and animals — that actually help grow the food. I see it as a trade. Agricultural corporations, on the other hand, want to reduce crop loss (and profit loss) as much as possible and long ago developed petroleum-based nutrients, pesticides, and herbicides in what they hoped would be both powerful insurance against crop loss and a boost in large-scale production. Flash forward 50 years, and we find ourselves suffering from this short-sighted view of cheating nature of her rightful 25 percent. We have contaminated our soil, our watersheds, the ocean, and our own internal ecosystems. Despite that, guess what today's net loss is? Approximately 25 percent.

When working with and growing cannabis it is important to consider how much yield is fair to ask of each plant. Is your approach coming from a grateful place or from greed? What are you (or your grower) willing to do to boost production? Will you continually dump nutrients into your plants to increase yield? A good rule of thumb, set by the industry for optimal indoor grows, is this: A 1,000-watt light covering approximately a 4-foot-by-4-foot canopy will yield 1 to 2 pounds of flowers per light. For 600-watt lights, you can expect about 1 pound of yield. Sure, growers will boast of 3 pounds of yield per light, and I'm sure you'll hear other tall tales. Sun-grown plants have a much larger range, from a few ounces per plant to 5 to 8 pounds! Environment plays a much bigger role in sun-grown plants. California, with its many days of sun and dry climate, yields on the higher end, while New England, with a shorter growing season, fewer sunny days, and a wetter environment, yields on the lower end.

We rely on the abundance and generosity of the plant world for so much: our food, our medicine, our clothing, our shelter. We would do well to continuously remember this and be grateful. Our gratitude increases our own health, the health and well-being of those around us, and the planet herself.

CHAPTER TWO

MEET THE CANNABIS PLANT

The cannabis plant had been living, surviving, reproducing, and just plain going about her business for some 60,000 years on Earth before we came along and wanted to use her for medicine, fiber, fuel, or food. In that time, the plant figured out a thing or two about what she needed and how to protect herself from predation by herbivores and insects. Cannabis also figured out how to protect her DNA from the sun's damaging ultraviolet rays, and she learned how to survive toxically high oxygen levels in the atmosphere and periods of life-threatening loss of water. Any medicinal benefit to us is a cumulative gift from all of our plant ancestors, who figured out how to live on this planet millions of years before we arrived.

The Origin and Migration of Cannabis

Cannabis ruderalis, the wild grandmother of the cannabis we know today, originated in the mountains of central Asia. This plant migrated and gave rise to the two species we now know as cannabis: *Cannabis indica* and *Cannabis sativa*. Today, different types of cannabis are classified by their leaf shape — broad or narrow — as well as the levels of CBD and THC present in the plant. In simplistic terms, CBD is the constituent in the plant that has a soothing effect on the body, and THC is the constituent that makes the body feel "high."

Migration

Cannabis sativa

Cannabis ruderalis

Cannabis indica

Cannabis indica chinesis

Cannabis indica indica

Cannabis indica afghanica

The Wild World of Nomenclature

Whether a cannabis plant is labeled as "hemp" or "drug/marijuana" has to do with its level of THC. "Hemp" is a term created by the government to classify any cannabis that contains 0.3 percent or less THC content (by dry weight). It is cannabis, but with such low levels of THC it cannot get you high. It is a standard imposed by the government, not biological or taxonomical nomenclature. Hemp is cannabis. There are two lineages of cannabis classified as "hemp," and they are labeled based on the shape of their leaves: narrow-leaf hemp (NLH) and broadleaf hemp (BLH). Hemp is grown for its fiber

IS CBD LEGAL?

There is confusion about the federal legality of CBD-related products. The 2018 Farm Bill legalized hemp, which is high-CBD cannabis with below 0.3 percent THC. So theoretically it is legal nationwide. But there's a catch: the U.S. Food and Drug Administration (FDA) says that because CBD is also an approved drug (called Epidiolex), the cannabinoid can't be considered a dietary supplement. Therefore, the companies that ship CBD products across state lines — an activity subject to FDA enforcement — may be violating the law. Yet even though the FDA has the authority to clamp down on CBD-related products and interstate commerce, it can choose not to do so. For the most part, at least for now, it's choosing not to. FDA enforcement action depends on, among other things, available resources, public attitudes, and the perceived threat to public health.

to make rope, clothing, paper, and building materials. It is also cultivated for food; its seeds are used to make hemp milk, hempseed oil,
and hemp nut butter. Hemp is also cultivated for CBD medicine. Most
industrial/agricultural hemp varieties have CBD levels of about 8 percent and require processing to create medicine. The 8 percent level
will continue to rise as breeders cultivate for plants higher in CBD.

The labels "drug," "recreational," and "medical" all denote cannabis flowers with a 0.3 percent or higher level of THC. These imposed
terms have underpinnings of judgment and a good-versus-bad paradigm; the culture already has a bias against the word "drug". What
determines the line between recreational and medical use? *Is there a
line?* An increase in quality of life is a common report when working
with cannabis; that makes cannabis a good therapeutic tool for any
chronic disease. In this book, unless citing scientific nomenclature,
I use the term "personal use" to indicate medical or recreational
use. The level of THC in personal-use cannabis usually ranges from
14 percent to upward of 30 percent. **But again, no matter what the
label, all of these flowers are cannabis.** And just like with hemp,
the lineages of personal-use cannabis are based on the shape of their
leaves: narrow-leaf drug (NLD) and broadleaf drug (BLD).

Up Close and Personal with the Two Species

Botanically — not common vernacular — there are two ways of
categorizing cannabis: as two species or three. For our discussion, we will use two species: *Cannabis sativa* and *Cannabis indica*
(*Cannabis indica* has three subspecies). If you step into a dispensary
or talk with your local grower, you will inevitably be asked if you are
looking for indica or sativa. People commonly say "sativa" for tall,
lanky plants that are energizing and heady and "indica" for short,
squat plants that are sedating or provide a body high. While common use of these terms is helpful for figuring out what someone is
looking for, it should not be confused with the taxonomical naming
by botanists. It can be confusing because what people call "sativa" is

Scientific name	Physical attribute	THC content	CBD content
Cannabis sativa	narrow-leaf hemp	low	high
Cannabis indica chinensis	broadleaf hemp	low	low
Cannabis indica indica	narrow-leaf drug	high	low
Cannabis indica afghanica	broadleaf drug	high	high

actually a subspecies of *Cannabis indica: Cannabis indica indica*. And for the record, you cannot determine a plant's sedative or energizing properties simply by looking at it.

Cannabis sativa
Narrow-leaf hemp (NLH); low THC, high CBD
Cannabis sativa evolved in the temperate Caucasus Mountains of western Eurasia. This plant is the descendent of cannabis low in THC and high in CBD; it is also called narrow-leaf hemp (NLH). Medicine makers wanting a high-CBD formula can utilize this species.

Cannabis indica and subspecies
Cannabis indica evolved in the Hengduan Mountains of Asia. *Cannabis indica* gave rise to three subspecies: *Cannabis indica chinensis*, *Cannabis indica indica*, and *Cannabis indica afghanica*.

CANNABIS INDICA CHINENSIS
Broadleaf hemp (BLH); low THC, low CBD
Cannabis indica chinensis originated in eastern Asia. It contains low levels of THC and low CBD, so it is bred and used chiefly for fiber and seeds. This subspecies would test low for the two main active constituents and would not be suitable for medicine.

CANNABIS INDICA INDICA
Narrow-leaf drug (NLD); high THC, low CBD
Cannabis indica indica originated in India. It is high in THC and low in CBD, and it is termed narrow-leaf drug (NLD). *Cannabis indica indica*

arrived in the United States in the nineteenth century. These strains, with high THC, were used early on as psychotropic medicines. The effects were described as "heady" and "energizing." In the recreational cannabis community, this plant is commonly called "sativa" today. This subspecies is typically a light green color, and strains can contain 20 percent or more THC. Over the years, cannabis breeders of NLD species have selected for intoxicating psychoactivity, creating highly psychotropic strains high in THC but very low in CBD. This subspecies is a good medicine when you desire higher THC, terpenes, and other constituents that have not been bred for but may exist in the plant.

CANNABIS INDICA AFGHANICA
Broadleaf drug (BLD); high THC, high CBD

Cannabis indica afghanica originated in Afghanistan. It is high in both THC and CBD and is termed broadleaf drug (BLD). It arrived in the United States in the 1970s. The plant is short, robust, and dark green. This subspecies can contain THC levels of 10 percent and above. It can also be high in CBD. Because CBD and THC both create the resin, plants were bred for their high resin content (to make hashish)

CULTIVARS, VARIETIES, AND STRAINS

So what about your favorite strains? Pineapple Trainwreck or Blueberry Muffin or Sour Diesel, perhaps? Where do they fit into all of this naming/taxonomy? What is a strain anyway? All of the varieties you see and hear about at dispensaries are descendants from an original lineage and are either *Cannabis indica indica* or *Cannabis indica afghanica*. They are subtle variations within a subspecies that contain different terpene profiles and cannabinoid content. (The terms "strains," "varieties," and "cultivars" are synonyms; we will use "cultivar.")

▲ A narrow-leaf plant in early flower

and their intoxicating psychoactivity. Selection favored CBD as well as THC, evidenced by the current levels of both constituents in BLD. This subspecies is commonly called "indica" in recreational and medical cannabis dispensary communities and is described as having more of a "body high" and sedative effect, giving the user what is commonly called "couch lock" (because you have difficulty getting off the couch).

To be clear, the botanical classification of *cannabis indica* and *cannabis sativa* is not the same as the common use of the terms "indica" and "sativa."

Anatomy of the Plant

Cannabis is dioecious: individual plants are usually either male or female. While both sexes produce flowers, male flowers have male reproductive parts that produce pollen while female flowers have female reproductive parts to receive the pollen. Pollinated female plants put a lot of energy into producing seeds, energy that unpollinated plants put into making big, juicy buds with the most potent medicinal potential. Therefore, growers try to remove male plants — along with their flowers and pollen — as soon as sex characteristics appear. Under stressful situations, however, such as extreme temperature or humidity, adverse nutrient or water levels, fluctuating light, or predation, a female cannabis plant *can* become monoecious and produce both male and female flowers in order to self-pollinate. This is a beautiful survival mechanism: when life is stressful and a plant senses it may have trouble reproducing, she can fertilize herself! (This is great for the wild cannabis plant but not so great for growers.) The plant's leaves have a classic palmate structure that resembles a hand. Early in the plant's life, the leaves are arranged oppositely on the stem; as the plant matures, the leaves arrange alternately.

For medicine, we use all parts of the plant that manufacture cannabinoids and terpenes. The structures that manufacture the cannabinoids and terpenes are called "trichomes." These sparkly "crystals" are found on the flowers, leaves, twigs, sepals, and perigonial bracts (specialized leaves). The flower itself is prized most highly because

Cannabis Plant Anatomy

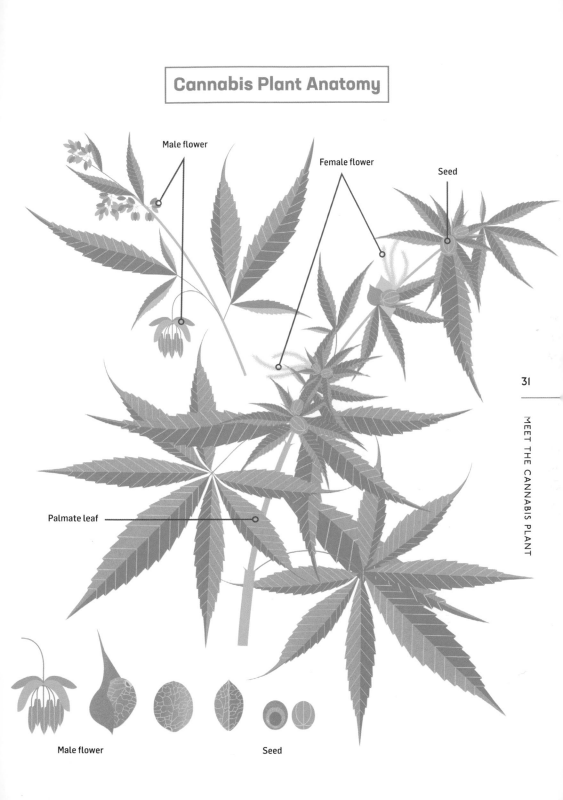

Male flower

Female flower

Seed

Palmate leaf

Male flower

Seed

it has the most potent trichomes and medicine, but all these parts contain trichomes and thus some medicine-making power. Growers will trim off all non-flower parts when getting ready for selling. These other trichome-containing plant parts are called "trim." While not nearly as potent as flowers, trim works fine for making medicine. It is not available for sale at dispensaries, but home growers may sell it at a substantially cheaper price than the flowers.

Cannabis is a genus of the *Cannabaceae* family, of which hops (*Humulus lupulus*) is the only other member. Interestingly, cannabis was formerly classified with nettles in the *Urticaceae* family, probably due to the presence of cystolithic and unicellular nonglandular trichomes on both plants. Cannabis and nettles both love to grow in nutrient-rich, well-drained soils with full sun, and both are jam-packed with nutrients, antioxidants, and minerals. Reports of the healing benefits of juicing cannabis leaves are all over the internet. Cannabis-leaf juice is nutrient rich but extremely bitter (as well as illegal in some states). Nettle leaves, on the other hand, are equally nutritious, tastier, cheaper, and legal. Herbalist Lisa Ganora calls cannabis "nettles with benefits."

Cultivar Chemicals

One of the ways plants communicate with us is through chemistry. When we smell a rose and feel joyful, we are interacting with the plant via chemicals. When we drink nettle tea and feel fortified, we are communicating with the plant through chemicals. The cannabis plant, with her thousands of variations of chemical profiles, is an apothecary in itself. Each cultivar has a different chemical profile with different healing properties. You could spend a lifetime getting to know just a few cultivars of cannabis and how they work under different conditions. Most people don't know what cultivar they are working with, and even fewer know the chemical breakdown of the particular cultivar (and therefore what the cultivar may be good for). But the better we understand the chemistry of the plant, the better our ability to make well-informed decisions about what cultivar to use, and what conditions it grows best under, when making and working with cannabis medicinally.

Yes, it's important to know the cultivars you are working with, and yes, it would be great to have them all tested for their chemical breakdown. But if you work with four or five particular strains and know their effects — stimulating or sedating or helping with pain — you know what conditions they will work for. Knowing the chemical profile of a plant *does not trump* your own experience with how the plant works.

The Magical World of Trichomes

Trichomes are resinous, hairlike glands. They are cannabinoid and terpene factories. They are easily seen with the naked eye; they look like sparkly jewels encrusting the flower. Trichomes cover cannabis flowers and bracts (the small leaves around the flower). They function much like the hair on our arms, which warms us and prevents water loss. Trichomes on the surface and undersides of the flower and bract trap warm and moist air on the surface of the leaf. In essence, they create an ideal micro-environment, protecting leaf, flower, and seeds from ultraviolet radiation in sunlight.

The cannabis plant makes six different types of trichomes. Only three of them produce resin, and we will focus on those. The first of these, called "capitate-stalked," looks like a tiny mushroom. They are the largest and most numerous, and they produce the most cannabinoid-rich resin. "Capitate-sessile" trichomes look like tiny bulbs; they have stalks, but they are very short. Finally, "bulbous" trichomes look like tiny lightbulbs atop little stalks.

In living plants, trichomes are sensitive and can easily break. They then release their resinous extract, which contains antibacterial and antifungal terpenes and can also trap and kill insects. Trichomes contain chemicals that are bitter, which deters herbivores. Cannabinoids, when released onto the surface of the flower, are necrotic to insects and the leaves, which degrade to expose the voluptuous female flower to the male pollen carried on the wind. That's right — when the wind blows, plant sperm is flying through the air! The female plant keeps growing the flower, extending and producing sticky resin to catch any of the sperm-containing pollen from the male flower. Once fertilization occurs, the female plant goes about the job of making offspring in the form of seeds.

It's important to note that trichomes in a dried plant are even more delicate. Anytime you handle a flower roughly by squeezing or dropping it, you break open trichomes, releasing the volatile terpenes and exposing the cannabinoids to oxygen. If you can smell the "cannabis smell," you've broken open some trichomes and released their terpenes. So to make the most potent medicine, handle the plant material as little as possible.

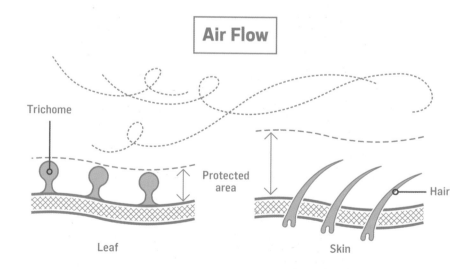

▲ Trichomes function much like the hair on our arms. They trap warm, moist air on the surface of the leaf, creating an ideal micro-environment to protect leaf, flower, and seeds from UV light.

Trichomes are resinous, hairlike glands that produce cannabinoids and terpenes. They look like sparkly jewels encrusting the flower and bracts (the small leaves around the flower). The cannabis plant makes six different types of trichomes; three of them produce resin: capitate-stalked, capitate-sessile, and bulbous.

▲ Capitate-stalked trichomes on the bract — these are the largest and most numerous, and they produce the most resin.

▲ **Capitate-sessile trichomes** on the underside of a leaf have stalks, but they are very short. Capitate-stalked are seen here with an observable stem.

▲ **Bulbous trichomes** on a leaf look like tiny lightbulbs atop little stalks. This is a closer view of capitate-stalked trichomes on a female flower.

▲ **Trichomes on female flower and bract.** Trichomes are delicate, and when broken they release a resinous extract that contains antibacterial, antifungal terpenes and necrotic cannabinoids.

▲ **Unpollinated female cannabis flowers** are called sinsemilla, "without seed." They are prized for medicinal use because all the plant's energy was invested in producing trichomes (which hold the beneficial chemicals) rather than seeds.

Sinsemilla Flowers

For medicinal purposes, we use unpollinated female cannabis flowers, called "sinsemilla," meaning "without seed." Sinsemilla flowers are prized because no energy was put into making seeds; instead it was invested into making more trichomes and the chemicals within them.

GROWING SINSEMILLA

It's worth noting that cannabis pollen can travel up to 300 kilometers (almost 200 miles). What does this mean for growers? If you are growing your plants outdoors and hope to have sinsemilla (or pure seed stock, where you know what strain the father is), you will need to make sure any neighbors growing any kind of cannabis remove all the male plants before they release pollen — unless they live both downwind and more than three miles away. Otherwise, their plants could be having unsupervised sex with yours!

Seeded Flowers

Cannabis ensures survival of her wisdom and lineage through seed production. Seeds are also an important food source for animals, who happily excrete them along with a nice dose of fertilizer, spreading the plants as they travel. Seeded flowers are useful, of course, for making sure you have something to plant the next season. And seeded flowers can be used for medicine, but they are less potent. In the world of commercial cannabis growers, seeded flowers are undesirable and bring the price of the flower down for two reasons. First, as mentioned, the plant puts energy into making seeds that could have instead made cannabinoid- and terpene-rich trichomes. Second, about 35 percent of the weight of a pollinated flower is seeds. Since seeds are useless to end users of cannabis, they won't pay as much for seedy flowers. The good news for medicine makers is that once the seeds have been removed, the flowers are only a little less potent than sinsemilla, and you may be able to buy seeded flowers for less — hopefully about 35 percent less.

▲ **Seeded flowers** can be used for medicine, but they are less potent.

▲ **About 35 percent** of the weight of a pollinated flower is seeds.

Banana
flower

▲ **A yellow stamen,** "banana," normally grows inside a male
pollen sac but sometimes they appear directly on female buds,
especially in times of stress. They are capable of pollinating
the female flowers and producing seeds.

Male Plants

Male plants contain 20 times less THC and CBD than female plants,
are typically taller than females, and flower one to three weeks ahead
of females. A single male flower can produce 350,000 pollen grains;
with hundreds of flowers, each plant can release hundreds of millions
of pollen grains into the wind and into female flowers. That's why
most growers remove their male plants as soon as it is possible to
sex them. Cutting and leaving male plants in the field will not protect
against unwanted fertilization of female flowers. The male flowers
must be removed from the field. Fun fact: In the midwestern United
States, 36 percent of late-summer pollen comes from wild cannabis!
So your hay fever might be an allergy to cannabis, not ragweed.

▲ This male flower is almost
ready to open and release pollen.

▲ Male plants are typically taller than females and contain
20 times less THC and CBD. Identifying a male plant early on
is crucial to prevent unwanted pollination.

▲ **The plants' preflowers** will indicate their sex during the vegetative phase. Female preflowers show two little white hairs, the pistils. Male preflowers show a small bulbous ball and no hairs.

▲ **This is a very early male flower.** If you can identify at this stage, you can decrease the risk of pollinating any female plants in the area.

The ability to identify a male plant early on is crucial to prevent unwanted pollination. Tests exist that can sex a plant by the leaf as early as a few weeks after germination. In states that regulate how many plants you can legally grow, it might be worth the money to test your plants early on to identify and remove males from the garden. If you don't want to pay for testing, there are guidelines that are useful in identifying male plants. First, if your plants are all of one cultivar, keep an eye out for the taller ones; they may be males. This is not a definitive indication of sex, just an early indicator. Second, the plants' preflowers will indicate their sex during the vegetative phase. Look for preflowers at the junction of the leaves and the main stem; they will appear approximately 10 weeks after germination. Female preflowers show two little white hairs, the pistils. Male preflowers show a small bulbous ball and no hairs. It can be difficult to spot the ball, but the presence of the two hairs signals a female.

Life Cycle of the Plant

Cannabis is typically grown as an annual, completing its life cycle in less than a year. A seed planted in the spring will grow into a tall plant through the summer and flower in the fall, producing more seeds to start the cycle again the following spring.

Germination and Seedling Growth

Life begins for the cannabis plant when the seed germinates, which takes 3 to 10 days after planting. One surefire way to germinate seeds is to sandwich them between layers of wet but wrung-out paper towels, seal it all inside a Ziploc bag, and store in a warm place. If you're starting hundreds of plants, it is probably best to sow directly into warm, moist soil that gets at least 16 to 18 hours of light. Seedlings should be protected from frost; if you live in the northeastern United States, start your plants indoors with supplemental light or wait until the danger of frost has passed. After germination, the tiny seedlings begin to grow rootlets branching out from the single root, and true leaves begin to form and grow.

▲ **In the fall,** female plants grow flowers sticky with resin in the hopes of catching male pollen carried by the wind. Fertilized female flowers put energy into producing seeds to continue the life cycle.

▲ **During the warm summer months,** cannabis plants can grow a few inches in a day!

Vegetative Growth

Vegetative growth can be astounding! On a hot summer day you can see growth of a few inches a day. The cannabis plant spends most of its vegetative growth phase accumulating bulk in its roots and stems and leafing out. Vegetative growth is triggered when the plant receives more than 12 hours of daylight per day. When the day length decreases to less than 12 hours, the plant transitions from vegetative growth to full-scale flower production. Where I live in upstate New York, outdoor plants begin their real growth spurt in June and, depending on the variety, begin flowering in August or early September.

Flowering Phase

After the long days of summer, when days shorten to less than 12 hours, the cannabis plant heads into its final life cycle and begins its work in perpetuating the species. In autumn, female plants continue to grow flowers sticky with resin in the hopes of catching male pollen (sperm) carried by the wind. Fertilized female flowers put energy into producing seeds, which, after fully developing, drop to the ground to become next year's expression.

The length of each cultivar's flowering period is genetically pre-determined, but all are roughly between 45 and 90 days. People living in northern latitudes should avoid cultivars with long flowering periods because frost can set in before flowering is complete. Reputable seed companies will provide this information (see the Resources section on page 249).

At some point during flowering, trichome color changes from clear to cloudy to golden as the plant matures. To see this change, you'll need a hand lens of at least 10× power. The optimum time to harvest is when the trichomes are mostly cloudy and beginning to turn golden; this is when they are the most packed with cannabinoids and terpenes. Clear trichomes are not fully matured, and medicine made from the plant at this stage can leave a person feeling jittery. Medicine made from a plant with trichomes that are fully golden, on the other hand, will often have more sedative effects.

Chemistry of the Trichome, Where the Magic Happens

If we dive into a trichome, we will find an ancestor molecule (precursor) named cannabigerolic acid (CBGA). The cannabis plant uses enzymes to convert CBGA into all the good stuff we want: THCA, CBDA, and CBCA. There is a finite amount of CBGA each cultivar of cannabis manufactures, and since every acid is made from CBGA, there is a maximum amount of the total number of acids the plant can produce. The enzymes needed to produce THCA or CBDA are genetically determined and of a finite amount as well. A good rule of thumb is a range of 20 to 30 percent total acid content in a plant. Therefore, we would not expect to see a plant with 25 percent THCA and 25 percent CBDA. There isn't enough CBGA to make all of the acids at that high a level. Cultivators until recently have focused on breeding for high levels of THC, but with the recent CBD craze there is a rush to develop plants with a high CBD profile and less than 0.3 percent THC (the federal legal limit of THC in a "hemp" plant).

Terpenes are also made in the trichomes. Their concentration depends on the amount of precursor molecules the plant starts with, the availability of conversion enzymes, and the environmental conditions the plant is living in. Practical data also suggests that terpene levels increase with increasing light levels, lower nutrient levels, and decreased soil water levels.

When the trichomes are intact, the chemical constituents are more stable and degrade slowly. Once the trichomes break apart, the chemicals degrade more quickly. You might notice dispensary workers carefully removing flowers from jars with large tweezers. I often wonder at the showmanship of this behavior because this is the last step in a process that is not as careful. You can be sure that harvesting, trimming, drying, bagging, and transporting to the dispensary have not been done with tweezer sensitivity. Every step offers a chance to degrade the potency of the medicine.

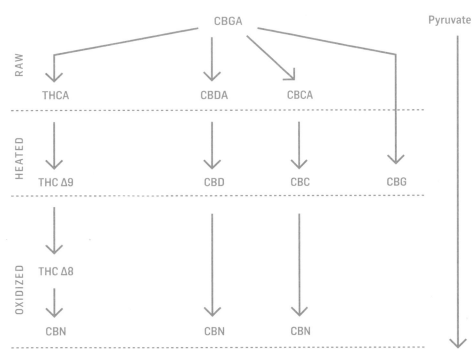

▲ Terpenes are an essential part of your medicine. If you want to retain terpenes, do not expose them to heat above 100°F (38°C). If you overheat cannabinoids or heat cannabinoids for too long, they will convert to CBN and cannot be reactivated — there's no coming back from CBN.

Chemical Constituents of Cannabis

Cannabis, like all plants, has hundreds (possibly thousands) of chemical constituents. Think of plant constituents as instruments playing in an orchestra; when they're all playing together, they can create a beautiful symphony. While it is interesting to understand what each instrument contributes on its own, the full power of the symphony is revealed only when the entire orchestra works together as one. Similarly, as we single out the constituents of cannabis here and discuss how each functions in the body, we must remember that they also work in coordination with one another within the plant and our body.

We use the many gifts of cannabis and her wisdom when we make medicine. We can broadly divide the more than 500 identified chemical constituents into four categories: acids, cannabinoids, terpenes, and flavonoids. We will spend most of our discussion on cannabinoids and terpenes because most of the research has been done with them.

Cannabis communicates with us and brings about physical healing when *her* chemical constituents bind with the receptors of *our* endocannabinoid system. Understanding the chemistry of cannabis will help us understand how she interacts with our chemistry, which will ultimately enable us to make better medicine.

Acids: THCA, CBDA, CBCA, THCVA, CBGA. These acid forms of cannabinoids are precursors to the active forms (created when the acid is removed through decarboxylation). Many have healing properties themselves.

Cannabinoids: THC, CBD, CBG, CBC, CBDV, THCV, CBN. These are the active forms of the above acids, the main constituents used in healing.

Terpenes: beta-myrcene, beta-caryophyllene, d-limonene, linalool, pulegone, eucalyptol, alpha-pinene, alpha-terpineol, terpineol, and many more! Terpenes are chemical compounds with repeating sequences of carbon and hydrogen. We recognize them in cannabis by their smell.

Flavonoids: apigenin, quercetin, cannaflavin-A, beta-sitosterol, luteolin, lutein, xanthophylls. Flavonoids are water-soluble pigments ranging in color from yellow to red to blue-purple. They are powerful antioxidants and anti-inflammatories.

Cannabinoids

Cannabinoids are the active constituents of the cannabis plant with the acid removed; they have beneficial healing functions in the body.

Trans-delta-9-tetrahydrocannabinol (THC)

THC was first isolated in 1964, prompting a search for the receptors that it binds to. The first receptor, named cannabinoid 1 (CB1), was discovered in 1988. In 1992, we discovered that our bodies manufacture a molecule similar to THC, which in turn led to the discovery of the endocannabinoid system, a system we continue to learn about.

THC IN THE PLANT

THC acts as an insecticide and fungicide for the cannabis plant. When it is released from the glands it also causes necrosis of the leaves around the flower at senescence (biological aging), which increases fertility of the female flowers by exposing more of their voluptuous stickiness to the wind and airborne pollen. THC levels increase with more light.

THC IN THE BODY

THC is highly lipophilic (lipid loving), so once absorbed, it travels in the blood bound to lipoproteins and albumin. It can either diffuse out of the bloodstream to interact with various target tissues and receptors (having effects on the body) or travel through the blood and enter the liver for detoxification. The liver detoxifies substances for excretion using a variety of enzymes, the most well-known of which are the cytochrome P450 enzymes (CYP450). THC is also stored in adipose (fat) tissue. It has a half-life of one to three days in blood.

THC has myriad functions in the body: it is an analgesic, anti-convulsant, antiemetic, anti-inflammatory, antinociceptive, antioxidant, antispasmodic, anxiogenic, and anxiolytic. It decreases angiogenesis, metastasis, and tumor growth. It's a euphoric, intoxicant, muscle relaxant, and neuroprotector. It induces apoptosis, potentiates chemotherapy meds, enhances opioid receptors, and is a sedative.

THC mimics endocannabinoids — our endogenous chemicals — and binds to our endocannabinoid receptors. Once there, it can modulate and regulate the immune system, ease inflammation and pain, and induce apoptosis (programmed cell death). It can help us relax and sleep. It can affect our cognition, emotional memory, and learning. It can ease nausea and cause hunger. When we understand how our own system functions, we can better understand how plant medicine works.

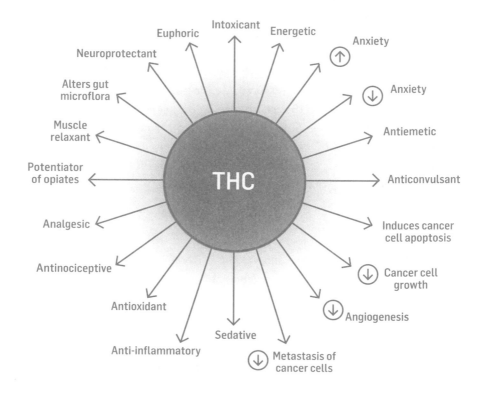

▲ THC plays a role in every body system, including the immune system and the digestive system. When we ingest THC from the cannabis plant, it mimics our own internal cannabinoids (AEA and 2AG).

THC and the Central Nervous System

At the level of the central nervous system (the brain and spinal cord), THC is energetic, euphoric, and intoxicating, causing spatial and temporal perception disorientation, motor discoordination, and short-term memory loss. It is psychoactive and a sedative. At higher concentrations and unopposed by CBD, THC increases anxiety (interestingly, it can also decrease anxiety, a paradox we'll discuss later).

THC and the Immune System

The immune system defends the body against foreign invaders (and against your own cells when they go rogue). It surveys the body for injury and signals for help. The immune system itself can be thought of as an intelligent system of communication: imagine tiny brains floating through all parts of your body. THC functions in the immune system to decrease inflammation and help protect the body against its own cells if they go rogue by inducing cancer-cell death, decreasing cancer-cell migration (metastasis), and throttling the growth of blood vessels to tumors (angiogenesis).

PHARMACOKINETICS AND PHARMACODYNAMICS OF THC

Because of its intoxicating properties, THC pharmacokinetics (the way a drug moves through the body) and pharmacodynamics (how a drug interacts with living structures) have been heavily studied; the nonintoxicating constituents of cannabis, with the exception of CBD, have not received nearly as much attention. CBD, because of its potent applications for healing (and its moneymaking potential for drug companies), is also now being heavily researched. The more studies we do, the more information we will have to make good decisions about medicine.

THC Absorption

THC most often reaches the bloodstream through inhalation or oral ingestion, but it can also be absorbed rectally and sublingually through blood vessels under the tongue. THC has limited transdermal absorption through the skin for entering the bloodstream but can act locally at the site of application.

INHALATION

When THC enters the bloodstream through the lungs, it travels first to the heart and then throughout the body, causing effects within 10 minutes. Only between 2 and 56 percent of inhaled THC makes it into the bloodstream (depending on the experience of the user and how much they are able to inhale), and a portion of that is broken down by the liver. Compared to other psychotropic drugs, THC is relatively slow-acting because its interaction with the body is rather complex. In general, the effects of inhaled THC in the brain are more delayed in onset than other inhaled psychotropic agents because the effects of THC are mediated further downstream than its primary binding site on the cannabinoid receptor.

ORAL CONSUMPTION

When we consume cannabis as an edible or tincture, THC enters the bloodstream through the gut lining. It then travels to the liver, where a portion is transformed — *presto chango!* — into the metabolite 11-hydroxy-THC, which is 5 to 10 times more potent than plain THC. Some 11-hydroxy-THC is broken down immediately, but if you've ingested more than the liver can quickly eliminate, or if the liver is busy processing something else, some 11-hydroxy-THC gets into the bloodstream, where it will travel around like inhaled THC and produce the same body effects. Since 11-hydroxy-THC is more powerful than inhaled THC, edibles can be quite potent!

Although edibles may be more potent, the effects take longer to appear, anywhere from 45 minutes to 3 hours, depending the contents of the stomach and how full it is, how well absorption occurs, and liver health. Tinctures are absorbed in 15 to 30 minutes, with the same mitigating factors as edibles. Peak concentration in the blood occurs in 1 to 8 hours. Bioavailability of orally ingested THC is 6 to 20 percent.

SUBLINGUAL ABSORPTION

There is lot of hype around sublingual/buccal sprays (see opposite page), but the evidence shows that they are pretty much comparable to tinctures in absorption. Sublingual absorption seems to have less variability, with bioavailability around 6 percent, and after initial absorption in the mouth there is no first-pass metabolism at the liver for removal from circulation.

FIRST-PASS METABOLISM

There's a lot of hype from companies who say that their product bypasses first-pass metabolism in the liver, but a little knowledge of physiology can help you sidestep the nonsense.

When we absorb cannabinoids (or anything) through our gut, the next stop is the liver for metabolizing, detoxifying, and filtering. But not everything is removed from circulation in this first pass. (How would pharmaceuticals work? And how would you get nutrients from food?) Some cannabinoids are removed from circulation, but most reenter the blood and travel to all the active sites. The blood then circulates back to the heart, to the lungs, back to the heart again, and back out to the body, all day long. Each time blood passes through the liver, cannabinoids are filtered out. The half-life of a chemical is determined by how long it takes the liver to remove half of it from circulation. THC has a half-life of one to three days, depending on how well the liver is working and how busy it is.

In sublingual absorption, it's true that when you absorb a substance through your mouth the next stop is not the liver. From the mouth, the blood travels to the heart, then to the lungs, and then back to the heart. From there it goes in many directions. Some goes to the limbs, some goes to the brain, and about a quarter goes to the liver. So while technically you bypass the liver the first time around, the stuff you spray in your mouth (and pay more for) ends up at the liver in about a minute anyway.

The tinctures you make and swallow have been, and will continue to be, potent and good medicine.

RECTAL

Bioavailability of THC via the rectum is two times higher than the oral route because of increased absorption rate and a decreased first-pass metabolism by the liver. It takes 2 to 8 hours for peak concentrations to occur in the blood.

TRANSDERMAL

Absorption of THC through the skin into the bloodstream is close to zero due to the lipophilicity of cannabinoids — they tend to accumulate in the skin without passing into the blood. Topical use of THC is beneficial, however, for *localized* applications.

Metabolic Breakdown of THC

Although the brain, small intestine, and lungs can all break down THC, the liver does most of the work. The liver breaks THC down into approximately 100 metabolites, but the majority is 11-hydroxy-THC, which is 5 to 10 times more psychoactive than THC because it so easily crosses the blood-brain barrier. "First-pass metabolism" is when THC is broken down by the liver *before* it can circulate and produce effects. Because anything absorbed by the intestines is first sent by the blood to the liver, orally administrated cannabinoids have less bioavailability than inhaled cannabinoids. The second phase of liver detoxification of THC results in the production of 11-nor-9-carboxy-THC or THC-COOH, which are both water soluble, not active in the body, and ready for the kidney to excrete.

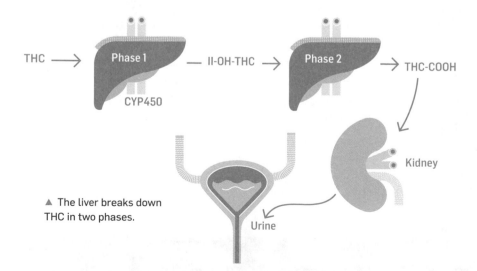

THC ⟶ Phase 1 — CYP450 — 11-OH-THC ⟶ Phase 2 ⟶ THC-COOH

Kidney

Urine

▲ The liver breaks down THC in two phases.

Excretion

About 80 to 90 percent of THC is excreted within five days; 65 percent is excreted in the feces as 11-OH-THC, and 20 percent is excreted in the urine as 11-nor-9-carboxy-THC or THC-COOH. THC-COOH can be detected in the urine 30 minutes after a single dose of inhaled cannabis.

TERMS DEFINED

THC is commonly characterized as psychoactive and CBD as nonpsychoactive, but for the record, CBD *is* psychoactive: it affects the mind and behavior. It is not an intoxicant or euphoriant per se, although some people do experience an uplift in mood when working with high-CBD strains of cannabis because CBD stimulates increased production of their own euphoria-inducing endocannabinoids.

The following terms are commonly applied to various drugs. Here's what each of them means.

Psychoactive: drugs that affect the mind or behavior

Psychotomimetic: drugs that mimic psychosis

Psychotropic: drugs that affect the mind, emotions, or behavior

Psychedelic: drugs that produce a profound sense of intensified sensory perceptions or hallucinations

Intoxicant: drugs that excite or elate

Euphoriant: drugs that induce a feeling of well-being or elation

Cannabidiol (CBD)

CBD is a relative newcomer on the cannabinoid scene. It was isolated in 1940 but wasn't identified until 1963. Because it is not an intoxicant or euphoriant, high-CBD plants have never been historically selected for the way high-THC plants have been. Nevertheless, CBD and THC are found in equal concentrations in *Cannabis indica afghanica* (BLD), likely because the selection pressures for making hashish favored plants with high resin content, and resin is rich in CBD. CBD is also the dominant cannabinoid in *Cannabis sativa* (NLH).

CBD IN THE PLANT

Cannabidiol protects the plant from ultraviolet (UV) radiation and acts as an animal and insect deterrent.

CBD IN THE BODY

The ability of CBD to bind to a vast array of targets underlies the myriad beneficial effects it has on the body. CBD's ability to bind to, or affect the binding properties of, different receptors is fundamental to its healing benefits.

In general, CBD modulates the endocannabinoid system. It is analgesic, antiemetic, anti-inflammatory, antioxidant, antipsychotic, and anxiolytic. It modulates the immune system and the effects of THC. It is neurogenic, neuroprotective, nonintoxicating, procognitive, psychoactive, and promotes neurite outgrowth and synaptic growth.

Now some specifics to amaze you . . .

Modulation of the Endocannabinoid System

CBD modulates the endocannabinoid system in two ways. First, it inhibits the reuptake of our own endocannabinoid, arachidonoyl ethanolamide (AEA), at the fatty acid–binding protein (FABP) transporter, thus keeping AEA circulating and active in the blood for longer. Second, CBD inhibits the enzyme FAAH, which breaks down AEA. Both mechanisms increase AEA levels and keep AEA in circulation longer, which is why some people experience mood elevation when they start consuming CBD-rich cannabis.

Modulation of the Effects of THC

CBD has a low affinity for cannabinoid (CB) receptors. It actually doesn't bind to the active site of a cannabinoid receptor (which are cell membrane receptors found throughout the body that are part of the endocannabinoid system). At very low concentrations, CBD actually acts opposite to what we would expect from an agonist molecule binding to a CB receptor. The action of CBD weakly binding to the CB receptor displaces some THC from binding and modulates the effects of THC, such as sedation, anxiety, and tachycardia. CBD weakly binding to the CB receptor also prevents some of the potent form of THC (that is, 11-hydroxy-THC) from binding to the receptor. The net effect is that less THC binds to CB1 and CB2 receptors, which has been shown to decrease psychosis, increase analgesia, increase the tolerability of THC, widen THC's window of activity, increase the volume of the hippocampus and amygdala, and potentiate the reduction of pain and inflammation. These effects were shown in a study of people who were given 200 mg of CBD per day for 10 weeks.

CBD and the Immune System

The effects of CBD on the immune system are nuanced and somewhat unexpected, but they fall into two categories: decreasing inflammation and fighting cancer. Cannabis is not an herb for combating colds, flu, or bacterial infections. Instead it systemically reduces inflammation (especially chronic) and fights cancer cells. CBD also increases production of natural killer cells (NKCs), which fight virus-infected cells and tumors.

CBD suppresses adaptive (learned) immune responses by quieting the activation of microglia (glia cells in the brain and spinal cord), decreasing proliferation and maturation of neutrophils (white blood cells), and inducing apoptosis (death) of monocytes and lymphocytes (other types of white blood cells) in the spleen and thymus. The suppression of the adaptive immune response is not a general suppression but rather an adaptive response to immune cells and microglia, increasing their production of inflammatory cytokines. This is the mechanism for shutting down inflammation. The production of inflammatory cytokines in the absence of an acute trigger is the definition of chronic inflammation; it is maladaptive and a contributing factor in chronic disease.

Cannabigerol (CBG)

Identified in 1964, CBG is decarboxylated CBGA, the precursor molecule for THC, CBD, and terpenes. CBG can be found in hemp fiber cultivars and some high-THC cultivars. CBG is still relatively new to the science community, so we have limited information on its benefits. CBG does look promising for fighting oral, prostate, and breast cancer. It may have antidepressant, antiemetic, antifungal, anti-inflammatory, antinausea, antiseptic, and neuroprotective properties. It decreases optic pressure, and increases AEA. CBG might also be a muscle relaxant and may promote skin cell proliferation.

Cannabichromene (CBC)

Cannabichromene acid (CBCA) is the acid form of CBC before decarboxylation. It is found in the early flowers of some cannabis cultivars and may protect the plant against fungal and bacterial infections during the flowering phase. CBC is an analgesic, an antidepressant, an anti-inflammatory, a sedative, and a potent inhibitor of AEA reuptake.

Cannabidivarin (CBDV)

CBDV is an analogue of CBD found in feral plant populations in India; most strains don't contain this constituent. We know that it is analgesic and increases production of 2AG, an endocannabinoid produced by the body. CBDV is currently being tested for its effects on type 2 diabetes, glaucoma, schizophrenia, seizures, and neonatal hypoxia.

Tetrahydrocannabivarin (THCV)

THCV is an analogue of THC with enough chemical differences to prevent its usefulness as a euphoriant or intoxicant. It is a minor cannabinoid found in cultivars from South Africa. It is an anticonvulsant, decreases body fat, lowers CB1- and CB2-induced hyperalgesia and inflammation, and at low doses reduces the effects of THC.

Cannabinol (CBN)

CBN is not found in live cannabis plants but is rather an oxidative by-product of THC or CBD, so the older a dried flower, resin, or oil is, the more CBN it will contain. CBN will naturally oxidize from THC and CBD in our medicines or dried flowers after approximately three years. CBN is not an intoxicant or euphoriant. It is anticonvulsant and anti-inflammatory, decreases keratinocyte overproliferation (as seen in psoriasis), and has sedative properties.

Acids

Cannabidiolic Acid (CBDA)

CBDA is the acid form of CBD. It is found in live plants and dried flowers that have not been decarboxylated. It is anti-inflammatory, analgesic, and binds the serotonin receptor 100 times more power-fully than CBD. The action at the serotonin receptor may also allevi-ate nausea and vomiting, anxiety, and depression.

Tetrahydrocannabinolic Acid (THCA)

THCA in the plant functions to protect against insect predation. When released, it causes apoptosis in the insect. In the human body, it is an anti-inflammatory, antiemetic, and analgesic.

Terpenes

Terpenes are classified chemically by the presence of repeating chains of carbon and hydrogen atoms (hydrocarbons).

Cannabis has more than 200 identified terpenes. As we identify them and learn more about their effects both in the plant itself and on our bodies, terpenes are becoming an increasingly important part of the constituent profiles being built for each strain of cannabis. Two different cannabis strains (say, an indica and a sativa) with exactly the same cannabinoid profile (for simplicity's sake, let's say they have the same THC and CBD ratios and percentages) might have extremely different healing capabilities. The difference comes from their varied terpene profiles.

Terpenes and Our Sense of Smell

One of the ways plants, through their terpenes, interact with us is through our sense of smell. Since terpenes are volatile and "smelly", we can interact with the plant. Olfaction, or our sense of smell, is processed in the oldest part of our brain and connects us to memories, to poison, to spoiled food, and lastly with spirit. The light terpene molecules released from the trichomes travel through the air into our nose and bind to olfactory receptors at the top of our nasal cavity. That activates a nerve impulse (electric current) that travels to the olfactory area of our brain and registers the scent of what we are smelling ("pine with a hint of lavender"). We understand indefinable things about our world, including plants, through our sense of smell. Plants communicate with each other and us through the olfactory universe of terpenes. We may not have the science to measure it yet, but that doesn't mean this communication is not happening.

I am an herbalist, not an aromatherapist, and my understanding of essential oils is limited. I do understand a simple yet profound and trustworthy diagnostic: if you like the smell of a plant, you're on the right track for working with it. If you don't like the smell, that particular cultivar might not be for you.

You or your client may or may not have access to labs for testing plant profiles. (As we've discussed, you can't always be sure the plant you have in front of you is the plant the person who gave it to you says it is.) You can "test" the plant yourself for basic characteristics, such as whether or not it is energizing, sedating, or anxiolytic, to know the basic criteria of the flowers of your apothecary. A further diagnostic for your clients (or yourself) would be to smell the flowers, holding the intention of what assistance you would like from the plant. The nose knows, and it can't lie.

Terpenes in the Plant

Terpenes and sesquiterpenes (heavy, extra-complicated terpenes), like cannabinoids, are made in the trichomes from acetyl CoA, the same precursor molecule of CBD, THC, and cannabigerolic acid. The number and types of terpenes produced in a plant are determined by genetics; terpene concentration also increases with sunlight and decreases with soil fertility. Plants make terpenes in part for protection against predation: produced by trichomes at the tops of the plants, bitter terpenes act as antifeedants against foragers. Terpenes and sesquiterpenes are also sticky and so can trap marauding insects before they cause too much damage. Finally, some terpenes offer protection against bacterial and fungal infections.

Terpenes in the Body

Some terpenes can bind to the CB2, GABA, NMDA, adenosine A2A, or 5-HT receptors. Like all essential oils, terpenes also bind to olfactory receptors in the nose. Terpenes are pharmacologically versatile. They can interact with receptors at the cell membrane or with ion channels within a muscle cell or neuron. They are able to change the fluidity of certain membranes, altering the permeability of the blood-brain barrier and the skin barrier. This change in membrane fluidity can increase the affinity of THC for the CB1 receptor and increase

TERPENES AND BIOAVAILABILITY

Terpenes are a whole medicinal apothecary unto themselves, and we can make more potent medicine by retaining them. Terpenes begin to evaporate at 70 to 100°F (21 to 38°C). To keep the terpenes in medicinal preparations biologically active, the medicine needs to be kept at room temperature or below. When inhaling cannabis, terpenes are contained in the smoke or vapor that you inhale. Any terpene concentration above 0.5 percent is considered biologically active and interesting to scientists. Terpenes in most inhaled cannabis flowers far surpass this level and are considered biologically active in the individual.

THC absorption across the blood-brain barrier, which could affect THC's analgesic and mood-altering properties.

Monoterpenes

The monoterpenes beta-myrcene, limonene, and pinene are all insect repellents and found in highest concentrations at the top of the cannabis plant. We will discuss a few, but not all, of the terpenes here.

BETA-MYRCENE

Beta-myrcene is the most abundant terpene in fresh cannabis; it is also found in hops and mango. It is analgesic, anti-inflammatory, antimutagenic, antiproliferative, and antipsychotic, and it increases the effects of THC in the brain. Cultivars with higher than 0.5 percent beta-myrcene tend to be sedating, whereas cultivars with less than 0.5 percent tend to be energizing.

LIMONENE

Limonene is the second most abundant terpene in fresh cannabis; it is also found in high amounts in citrus fruits. It is antibacterial, anticonvulsant, antidepressant, antifungal, and anxiolytic, and it decreases gastroesophageal reflux and serum cortisol (helping with stress). It is an immunostimulant and induces apoptosis in breast cancer cells.

PINENE

Pinene is the most abundant terpene in the plant world. It acts as a bronchodilator, increasing the absorption of other constituents when inhaled. It also is antibiotic, is anti-inflammatory, promotes focus and memory, and increases energy and sense of self-satisfaction. Cultivars with high amounts of pinene would be helpful for focus and concentration.

LINALOOL

Linalool is responsible for the floral scent of lavender. It is an analgesic; has anticancer, anticonvulsant, antipsychotic, and anxiolytic properties; can be used as a local anesthetic or sedative; and supports sleep.

TERPINOLENE

Terpinolene is found in lilac, apple, cumin, tea tree, lemon, sage, marjoram, rosemary, and pine. Some strains can be as much as 53 percent terpinolene. It has antibacterial, anticancer, antifungal, and antioxidant properties and is a sedative.

Sesquiterpenes

Sesquiterpenes are heavier molecules. They are less volatile than terpenes and have stronger odors. They are also anti-inflammatory and have bactericidal properties.

CARYOPHYLLENE

Caryophyllene is the most abundant sesquiterpene in the dried cannabis flower. It has a sweet and woody smell and is also found in black pepper, cinnamon, and clove. It is highly lipophilic, meaning it crosses the blood-brain barrier easily and is highly available to the brain. It also acts as an analgesic, anti-addictive, antibacterial, antidepressant, anti-inflammatory, antioxidant, antiproliferative, anti-seizure, antispasmodic, and anxiolytic.

ALPHA HUMULENE

Humulene is found in clove, basil, and hops. It carries a subtle earthy, woody aroma with spicy herbal notes. It is analgesic, antibacterial, anti-inflammatory, and antitumor, and it decreases appetite.

Flavonoids

Flavonoids are commonly known as plant pigments and protect the plants from damaging ultraviolet radiation. The pigments in other flowers (cannabis is wind pollinated) also attract pollinating birds and insects, carrying on the dance of reproduction. Do you like the purple color of your "blueberry muffin" cannabis strain? You can thank the flavonoids anthoxanthin and anthocyanin. Flavonoids are lost upon heating, so fresh preparations are necessary to retain the benefits of flavonoids in medicine. Try juicing, tinctures, or unheated oils.

Cannabis makes approximately 20 flavonoids, including apigenin, beta-sitosterol, cannaflavin (unique to cannabis), kaempferol, lutein, luteolin, orientin, quercetin, silymarin (also found in milk thistle), vitexin, and xanthophylls — all with anti-inflammatory, antifungal, antiviral, antioxidant, and anticancer potential. We use flavonoids as antioxidants and anti-inflammatory agents and to help prevent diseases associated with chronic inflammation. Cannaflavin A has anti-inflammatory properties and is 30 times more effective than aspirin at decreasing the inflammatory prostaglandin PGE2.

Chlorophyll

Chlorophyll, another pigment, gives plants their vibrant green color just like hemoglobin gives our blood its rich red color. The two molecules are almost exactly identical except for the element the molecule is built around. Hemoglobin is built around iron, while chlorophyll is built around magnesium. The gorgeous color in your infused oils and tinctures and teas comes from the extracted chlorophyll from the plant into your tea. Chlorophyll aids in the production of our red blood cells; it is anti-inflammatory, antioxidant, and antimutagenic, and it protects us from carcinogens.

CHAPTER THREE

THE ENDO-CANNABINOID SYSTEM

The endocannabinoid system (ECS) has only recently begun appearing in anatomy and biology textbooks. We are just beginning to understand the ECS, how it functions, and how it benefits us. Our understanding of the ECS is like a mosaic we are assembling over time. As more of the pieces come together, our understanding grows of how the ECS and cannabis fit into our health and well-being. The mosaic is filling in quite rapidly, considering we didn't know this system existed 25 years ago.

The ECS sets the baseline tone of well-being in which an organism (you) operates. When healthy, the ECS creates a background signal that all is well and quietly hums as the other systems go about their daily jobs, including the loud and fast branches of the nervous system, the quiet and steady heartbeat, the ever-aware immune system, the slow and steady endocrine system, the intelligent and interactive digestive system, and the hot-blooded reproductive system. Think of the ECS as the wizard behind the curtain, keeping you calm and centered; it is the basal system behind all other systems.

This integral and integrated system influences every system in the body, in both health and disease. It is so important that it has been evolutionarily conserved throughout all three kingdoms of life: plant, animal, and fungi (insects are the only major phyla who do not have the ability to interact with cannabinoids). Despite the fact that the ECS is so ancient, it was discovered only in the 1990s.

The ECS is physiological like the immune system, not anatomical like the cardiovascular or digestive system. While we can trace a blood vessel through the organs of the cardiovascular system or track a bite of food through the digestive system, the endocannabinoid system is the collective functioning of all cells in the body that can either make endocannabinoids or have receptors sensitive to endocannabinoids. At a cellular level, the endocannabinoid system operates like the endocrine system: chemicals are made by cells, then travel to and bind with receptors on other cells to produce a certain outcome.

The chemicals made by the ECS are called endocannabinoids. "Endo" means "within" or "internal." Our *endo*cannabinoid system is our own internal cannabis system. Cannabinoids are chemicals produced by cannabis that can bind to ECS receptors. If we can learn how endocannabinoid molecules bind to receptors, and learn about the actions such binding stimulates, we can understand how THC, CBD, and other chemical constituents of cannabis affect our bodies.

The bodily effects of consuming the cannabis plant are caused by cannabis constituents interacting with our endocannabinoid system. Plant molecules enter our bodies through our nose, lungs, skin, or digestive system. They then travel to and bind with cannabis receptors, an action that causes myriad physical reactions and sensations. Cannabis receptors permeate nearly every tissue in the body, so cannabis can interface with and cause changes through the whole body. No other plant shares such an intimate chemical relationship with us.

The ECS is complex, like a branching, weblike mycelial network. When looking at the ECS, we must think beyond the one chemical/one receptor/one reaction model; the ECS is a multifaceted, interconnected bodywide cellular network. As we understand this system more and more, we can develop truly integrated methods of approaching health and healing.

Discovery of the ECS

Dr. Raphael Mechoulam is credited as the "grandfather" of the endocannabinoid system. An Israeli organic chemist and head of the medicinal chemistry lab at the Hebrew University of Jerusalem in Israel, Mechoulam was among the first to isolate specific cannabinoids (including the main psychoactive constituent of cannabis, delta 9-THC) and elucidate their structures. He discovered the first cannabinoid receptor, called CB1, in 1988 when he was trying to understand how THC worked in the body. He realized that when people consumed cannabis, THC bound to the CB1 receptors in our bodies, causing the blissful feeling of well-being. He wondered if our bodies made a similar chemical of its own that would also bind to the receptor. In 1992, he discovered arachidonoyl ethanolamide (AEA), the body's own version of THC. AEA was nicknamed "anandamide" by Raphael's team because *ananda* means "bliss" in Sanskrit. AEA is our very own bliss molecule!

Mechoulam dubbed the ECS a "global protection system" because the general strategy of the ECS is to help cells, tissues, and organs reestablish a steady state of balance after acute or chronic disruptions, a condition called an "allostatic state."

System of Safety and Well-Being

Our baseline feeling of safety and well-being is the product of our endocannabinoid system. When this system functions properly, our nervous, endocrine, and immune systems are in safe mode. They're not sounding the alarm but simply going about their jobs; all is well. Healing, repair, digestion, and reproduction function optimally and we go about our day. Our emotional and psychological states also reflect our "all is well" status, and we experience a sense of well-being and curiosity. This baseline state is the most energy-efficient mode of operating. Animals operate in this mode except when they are chasing or being chased. A problem we humans often face is that as far as our nervous system is concerned, we feel like we are chasing or being chased all day long. It doesn't matter if it's actually happening or not — our brains treat it all as real (go ahead and check your pulse while watching a scary movie).

To get a feel for the ECS, let's go on a little journey to Grand-mother's house. It's been a long and tiring day, and the sun is going down. A cold wind kicks up, and a driving rain seeps under your collar as you trudge through the gray evening. The sidewalk is crumbling, and a car whips through a puddle and splashes you. Finally you make it to the house. You open the door and walk into a burst of warmth from the fireplace. You can smell a steaming mug of hot chocolate waiting by the fire. You peel off layers of wet outer clothing and sit in a soft chair. Grandma hands you a mug of hot chocolate, drops in a few marshmallows, and looks at you with love. "How are you doing today, my dear?" You are at home; you are safe.

This is a lovely simplification, yes, but it illustrates the complexity of our autonomic nervous system. This system has three branches: the sympathetic (which governs our "fight-or-flight" response), the parasympathetic (which handles rest and digestion), and the enteric (which innervates the gut). These three branches work together to operate all unconscious bodily activities: heartbeat, peristalsis of the digestive system, breathing, pupil dilation, and so on. Thankfully, these functions run automatically, below our conscious level of awareness. The endocannabinoid system operates behind the autonomic nervous system and sets the baseline tone.

If I were to interview a person to determine the health of their ECS, I would begin by asking questions like:

◆ Do you feel generally safe and secure?

◆ Are you able to explore new things?

◆ Are you able to relax?

◆ Are you curious about learning new
  things and taking in new information?

◆ How are you sleeping?

We should be able to answer positively to all of these questions. This state of well-being is our birthright — the baseline we share with all creatures.

Overarching Functions of the ECS

In healthy human bodies, the ECS activates in response to injury. Endocannabinoids are released in response to cellular stress in order to reestablish allostasis. The ECS interacts with nearly every cell of the body, and its proper functioning does not require a top-down hierarchy like the nervous system or a centralized chemical factory like the endocrine system. Almost each one of the 50 trillion cells of your body can affect its own health and well-being. We also know that a chronically stressed ECS can become dysfunctional, contributing to pathological conditions.

Before we dive into the biochemical and cellular functions of the endocannabinoid system, let's address the general functions of the ECS. These include allostasis, protection, perception of well-being, cognition and learning, emotions, feeding, and immunity; the ECS can also act as a reward system. Then we can fill in the mosaic of the ECS at the molecular, receptor, and chemical levels. Understanding how our own internal cannabis system functions will help us understand how cannabis works in us and brings about healing.

Allostasis

Our bodies are able to coordinate 50 trillion cells to keep us healthy and maintain proper body temperature, blood glucose and oxygen levels, heart rate, and breathing rate. When we can't maintain these parameters, we experience disease or even death. Allostasis is the maintenance of stability through physiological or behavioral change. To reach allostasis, we regulate our body functions as needed to adjust to new or changing environments: when we need to move fast, for example, our heart rate must go up to supply nutrients and oxygen to the muscles that need it. Allostasis (sometimes called homeostasis) is the body's ability to maintain equilibrium in an ever-changing environment. The endocannabinoid system works to bring about and maintain balance.

Protection

The ECS works to reduce inflammation, which is present in all major diseases. The ECS also helps protect against uncontrolled and unregulated cell growth, or cancer (for more on this see page 218). Pain, which keeps us from hurting ourselves or doing tissue damage, is also regulated by the ECS.

Perception of Well-Being

The ECS, with its endocannabinoid "bliss" molecules, signals that all is well. The system also interfaces with other various neurotransmitters (serotonin, dopamine, norepinephrine, and GABA) within the nervous system that are linked to the perception of well-being. A healthy, functioning ECS offers a baseline feeling of safety that allows us to sleep.

Cognition and Learning

The ECS functions within the brain to assist with learning, memory, and expanding consciousness.

Emotions

Having an overall perception of safety is intricately linked to our feelings of well-being. A healthy ECS decreases anxiety and increases curiosity. When we have a healthy ECS, we are less emotionally reactive, which also has a neurological benefit: we forget to be afraid of situations that are no longer dangerous. People who suffer from PTSD aren't able to do this. A healthy ECS is also linked to emotional memory and enables us to relax.

Feeding

The ECS controls feeding behaviors by stimulating our appetite (this is why cannabis gives us the munchies). The ECS also has a hand in controlling cell metabolism and regulating blood sugar levels. Nausea and vomiting are regulated by the ECS in both the brain and digestive lining.

Immunity

The ECS modulates the immune system by regulating the immune cells themselves (B, T, and natural killer cells). The ECS, like the immune system, also works to decrease inflammation.

Reward System

Nature gets us to do things like procreate or eat by rewarding us with good feelings via neurotransmitters that hit pleasure centers in our brain and send a signal: Do that again! The endocannabinoid system is directly linked to this neural reward system and plays a role in regulating it.

Functions of the ECS by Body System

Before we head into the functions of the ECS by body system, it is helpful to have some of the basic chemical functioning of the system under our belts. We will discuss the chemistry in depth on page 81. The ECS is made up of (1) chemicals that bind to receptors (for our purposes here, they are AEA and 2AG), (2) receptors that these chemicals bind to (CB1 and CB2), (3) transport molecules for endocannabinoids, and (4) enzymes that create or destroy endocannabinoids. In general, endocannabinoids work by binding to receptors on target cells that have receptors for them, activating the cell to carry out a specific function.

When we understand how a healthy ECS interacts with other body systems, we can begin to understand how the chemicals found in cannabis can help the body achieve a healthier state.

In Utero

The ECS plays an essential role in early embryonic and prenatal brain development (CB1 receptors have been seen in two-day-old mouse embryos). From our very beginnings, the endocannabinoid system is guiding us home: fertilized human eggs prefer to implant where the uterine lining produces high levels of AEA. Our bliss molecules (AEA) become the chemical flare for the fertilized egg. We implant into bliss! The ECS is deeply involved in many aspects of neural development: proliferation and differentiation of neuronal stem cells, neurodevelopment, creation of functional and effective synapses, orchestration of axonal migration and connection, and modulation of excitatory and inhibitory synaptic transmission in postnatal brain and spinal cord.

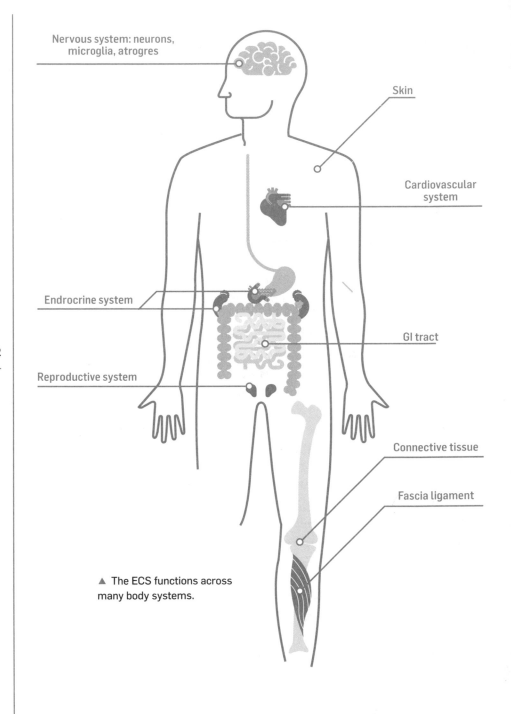

Nervous system: neurons, microglia, atrogres

Skin

Cardiovascular system

Endrocrine system

GI tract

Reproductive system

Connective tissue

Fascia ligament

▲ The ECS functions across many body systems.

Nervous System

The ECS functions within the nervous system to regulate neuro-genesis, neural protection, neural plasticity, autonomic tone, stress response, and pain.

NEUROGENESIS

The endocannabinoid system is responsible for stimulating growth of new neurons.

NEURAL PROTECTION

When neural tissue (or any tissue really) experiences chemical or physical trauma, it produces AEA and 2AG for protection and forms more CB2 receptors. This enables endocannabinoids to bind to the receptors, which decreases inflammation and prevents neuron damage. Glutamate, an excitatory neurotransmitter, can excite a neuron literally to death. When neurons are damaged by head trauma or a stroke, glu-tamate production can become unmitigated. The ECS mitigates gluta-mate excitotoxicity and decreases seizure activity in the brain.

NEURAL PLASTICITY

Neuroplasticity is the brain's ability to reorganize itself by forming new neural connections throughout life. Neuroplasticity enables neurons in the brain to compensate for injury and disease and respond to new situations or changes in the environment. Endocannabinoids stimulate the production of neurons (neurogenesis) in some regions of the brain.

AUTONOMIC NERVOUS SYSTEM

The ECS interacts with all branches of the autonomic nervous system via the countless CB1 receptors on neurons and the neuroglia sur-rounding them. Within the sympathetic system, endocannabinoids bind to CB1 receptors to slow the release of norepinephrine, a stimu-lating hormone and neurotransmitter. They also decrease sympathetic mediated pain and modulate the hypothalamic-pituitary-adrenal axis, which is the major pathway for the stress response. Within the para-sympathetic nervous system, endocannabinoids bind to receptors to decrease vomiting.

STRESS RESPONSE

A healthy ECS is key to the body's ability to adapt to stress and to recover after a stressful event. A healthy ECS helps us mount an appropriate stress response when needed and prevents us from overreacting when it's not. The ECS also initiates the cessation of the stress response.

Our fight-or-flight stress response is a beautifully orchestrated set of mechanisms that helped our ancestors survive in a tooth-and-claw world. Those who did not have a hypervigilant system didn't survive an attack, didn't reproduce, and are no one's ancestors. Hypervigilance serves us when we are in danger, but it is counterproductive and unhealthy when we are not. The problem is that once our brain has identified a danger, the rest of the body responds. Trying to stop that cascade goes against millions of years of selected-for biology. Given what we know about unmanaged stress causing major disease, the mechanism of the fight-or flight response (and how to mitigate it) warrants further exploration.

How is it possible on one day a loud noise makes you jump and on another day you hardly notice it? The answer is we are able to set and choose the threshold for response. This occurs in the region of the brain known as the amygdala. Alarm reactions begin here. Under stressful conditions, this region compares all incoming information to previous dangerous experiences. If they match closely enough, the amygdala perceives a threat and sounds the alarm. The threshold for sounding the alarm depends on many factors, but an important one is how much influence information from elsewhere has on the amygdala. Information from our higher, conscious brain can override an excited amygdala, calming it down before it goes on full alert. A healthy hippocampus, another brain structure that in part inhibits excitatory responses, can feed the amygdala information without tagging it as dangerous. Ironically, the stress hormone cortisol can shrink the hippocampus; the good news is that when cortisol levels drop, the hippocampus grows. Endocannabinoids binding to CB1 receptors help the hippocampus maintain its size and connections.

The region able to actually override the amygdala is in the higher brain, the prefrontal cortex (affectionately called the "angel lobes," or the lobes of compassion). Under stressful conditions the amygdala reacts to potential danger before the cortex can process the information, but with practice (creating strong synaptic connections through mindfulness and meditation) the prefrontal cortex can help keep the amygdala from hijacking your emotions. Meditation during non-stressful times strengthens the pathways that will help you keep cool during stressful times.

The heart can also override the amygdala. The heart is neurologically hardwired to every region of the brain, including the amygdala; and the brain can't shut these signals down. Techniques that focus on breathing and bringing attention and awareness to the heart can help soothe the amygdala during stressful times.

If we constantly perceive the world as threatening, our nervous system remains on high alert. We are neurologically primed to see danger everywhere. Our cortisol levels rise, shrinking the hippocampus and further diminishing our ability to feel secure, and a feedback loop is created.

Corticotropin-releasing hormone (CRH) is a chemical released by the amygdala with one function: to lower AEA levels, allowing or continuing the stress response. Chronic activation by the amygdala lowers AEA levels, which in turn leads to an easier triggering of the fight-or-flight response.

A healthy endocannabinoid system helps to decrease acute stress responses. Healthy AEA levels tamp down stress response in the amygdala, where we perceive threats. When we need to shut down a stress response, rising cortisol levels stimulate an increase in 2AG, diminishing the stress response, extinguishing fear, and enabling us to recover from the stressful event. Endocannabinoid deficiency contributes both to easier activation of the stress response and a diminished ability to recover.

If the ECS is deficient, our ability to perceive threat in a healthy way is diminished; we will overreact and the heightened stress response will continue.

Pain

The endocannabinoid system works to decrease pain on multiple levels, beginning at the point of injury all the way up to the perception of pain in the brain. We will more fully discuss the pain pathway

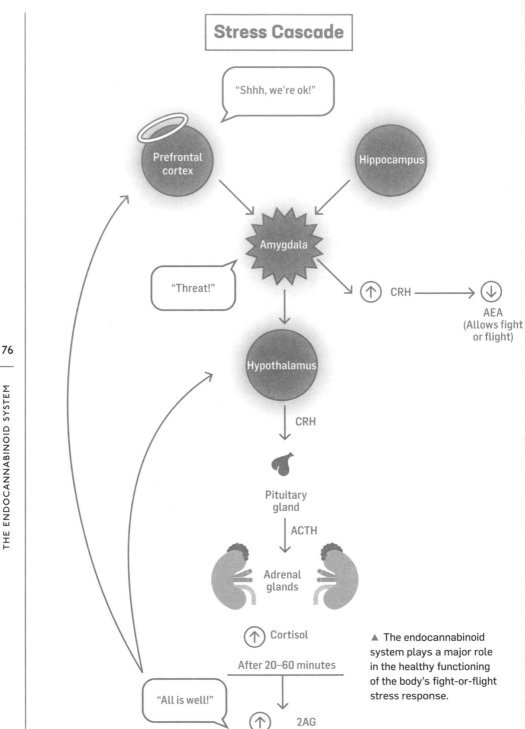

▲ The endocannabinoid system plays a major role in the healthy functioning of the body's fight-or-flight stress response.

in the chapter on conditions (page 164), but briefly, the ECS acts to modulate pain at the injury site, nociceptor (pain receptor), immune cells, and central nervous system.

INJURY SITE

When tissue is injured, cells produce chemicals (including histamine, prostaglandins, bradykinin, norepinephrine, substance P, hydrogen ions, potassium ions, leukotrienes, and adenosine triphosphate) that signal injury and cause the perception of pain. The endocannabinoids 2AG and AEA decrease the production of these chemicals, lessening pain.

NOCICEPTORS

When endocannabinoids bind to CB1 receptors on nociceptors (specialized pain receptors), they reduce their rate of firing; fewer pain signals sent to the brain means you feel less pain.

IMMUNE CELLS

When endocannabinoids bind to CB2 receptors on mast cells and macrophages, it causes these immune cells to slow their production of activating and sensitizing chemicals binding to nociceptors. This again decreases the signal for pain. The body in an effort to regulate pain will increase production of CB2 receptors after an injury to help decrease pain and inflammation as an allostatic function.

CENTRAL NERVOUS SYSTEM

There are several mechanisms within the central nervous system through which the ECS helps modulate pain. We will discuss these in detail in the chapter on conditions (see page 164).

PAIN WITH SPASTICITY

Some conditions are associated with muscle spasms, like spinal cord injury, multiple sclerosis, cerebral palsy, stroke, brain or head trauma, amyotrophic lateral sclerosis, or hereditary spastic paraplegia. Spasticity sometimes occurs when motor neurons can't stop firing, always telling a muscle to contract. Endocannabinoids help to increase the neurotransmitter GABA, which inhibits nerve impulses to the muscle, resulting in relaxing the muscle.

Cardiovascular System

CB1 and CB2 receptors are found all over the cardiovascular system. Under normal conditions, the ECS plays a limited role in cardiac function, but under pathological conditions such as septic shock, hemorrhagic shock, myocardial infarction, advanced liver cirrhosis, or doxorubicin-induced heart failure, the ECS lowers blood pressure by decreasing heart contractility. Endocannabinoids binding to CB1 and TRPV1 receptors on vascular tissue causes vasodilation, which also lowers blood pressure. Research on rats shows that the ECS also protects against myocardial tissue ischemia.

Immune System

The endocannabinoid system functions extensively within the immune system by modulating immune cell function and lowering inflammation. Endocannabinoids bind to CB2 receptors on helper T cells to decrease inflammatory cytokines and increase anti-inflammatory cytokines. The same mechanism also increases T cells, B cells, and natural killer cells.

ANTINEOPLASTIC EFFECTS (ANTICANCER)

Endocannabinoids, CB1, and CB2 receptor numbers are upregulated (increased in number) in tumor tissue, making tumors more sensitive to the effects of endocannabinoids. While chemotherapy kills all rapidly dividing cells, both healthy and cancerous, endocannabinoids target only cancer cells; their effects include inducing apoptosis and autophagy, suppressing angiogenesis, and inhibiting cancer migration. We will discuss cancer more fully in the conditions chapter (see page 164).

Apoptosis is a genetically directed process of cell self-destruction; it is a normal and controlled part of an organism's growth and essential to health. The binding of endocannabinoids to the CB1 receptor stimulates apoptosis.

Autophagy is the body's way of removing damaged cells and replacing them with new ones. It is a natural regenerative process that occurs at a cellular level, reducing the likelihood of contracting some diseases as well as prolonging life span. The binding of endocannabinoids to cannabinoid receptors causes enzymes within the cell to digest the cellular contents. This is especially important in immune cells that have engulfed foreign cells or cancerous cells.

Suppression of Tumor Angiogenesis
Angiogenesis occurs when tumor cells secrete chemicals that stimulate the growth of blood vessels. The binding of endocannabinoids to the cannabinoid receptors on tumor cells stops the production of these chemicals and shuts off the tumor's blood supply, starving it to death.

Inhibition of Cancer Migration
Sometimes cancer cells migrate (metastasize) to other areas of the body to find a better blood supply. The binding of endocannabinoids to the cannabinoid receptors on the tumor cells prevents them from migrating.

Connective Tissue
Connective tissue includes blood, bone, fascia, ligaments, tendons, and cartilage, together the most abundant category of tissue in our body. The endocannabinoid system interacts with and has beneficial effects on them all.

BONE
Bone tissue is made up of osteoblasts (bone builders) and osteoclasts (bone degraders). Both types of cells produce AEA and 2AG and express CB2 receptors. When AEA and 2AG bind to receptors on osteoclasts, bone breakdown is slowed. When AEA and 2AG bind to receptors on osteoblasts, bone building resumes.

AEA and 2AG binding to CB1 receptors in nerves close to the osteoblasts prevents the release of norepinephrine. Norepinephrine, a neurotransmitter of the stress response, decreases bone formation. When you are chronically stressed, your body doesn't want to allocate resources to building bone. It would rather allocate resources toward escaping from wild animals.

FASCIA AND CARTILAGE

The cells that make up fascia, tendons, and ligaments are called fibroblasts. Cells that make cartilage are called chondroblasts. Both express CB1 and CB2 receptors as well as endocannabinoid metabolic enzymes. It has been shown that cannabinoids (and presumably endocannabinoids) modulate fascial remodeling, prevent cartilage destruction, and decrease connective tissue inflammation. There is much more to be understood about the fascial system itself and the role that endocannabinoids play in it.

Digestive System

Both CB1 and CB2 receptors are found in the digestive system. CB1 receptors are found primarily within the enteric nervous system (the branch of the nervous system coordinating digestion), while CB2 receptors are found in the digestive tissue itself. Hunger and cell metabolism are both regulated by the ECS via regulation of the hormones ghrelin, leptin, orexin, and adiponectin.

When CB1 receptors in the enteric nervous system are bound by 2AG and AEA, digestion slows, gastric acid secretions decrease, the esophageal sphincter relaxes, intestinal motility slows, gastric motility decreases, and gastric emptying is delayed. CB2 receptors are likely involved in the inhibition of inflammation and visceral pain. In pathological conditions such as obesity, adipocytes overproduce CB1 receptors and endocannabinoids, resulting in a positive feedback loop leading to hunger, increased eating, and ultimately metabolic syndrome.

Hepatic Function

Healthy livers have relatively low levels of CB1 and CB2 receptors and low levels of AEA and 2AG. There are also low levels of the degradation enzymes FAAH and MAGL. All that changes when the liver is injured. Upon injury, CB2 receptor and 2AG levels increase within the liver.

When CB1 receptors are bound by endocannabinoids, fibrogenesis occurs in the liver; when CB2 receptors are activated, fibrosis is prevented. Because CB2 receptor activation is proven to decrease inflammation in connective tissue, it follows that it would also prevent fibrosis in the liver.

Psyche Health

The ECS sets an overall tone of safety and well-being in the body. This enables animals (including us) to learn new things, express curiosity, rest, relax, eat, and evolve. The ECS mechanism that creates this state is endocannabinoids binding to CB1, CB2, and TRPV1 receptors. This has the following effects (which we will discuss in depth in the chapter on conditions, page 164):

◆ Uncouples the stimulus-response reaction

◆ Allows painful memories to fade

◆ Inhibits the ability to reactivate painful memories

◆ Blocks anxiety related to chronic unpredictable stress

Chemical Messengers of the ECS

Endocannabinoids begin as a precursor molecule in our cell membranes. The precursor molecule migrates into the cell to interact with synthesizing enzymes that convert it into one of our body's different endocannabinoids. These molecules are then transported out of the cell to bind with receptors nearby. When they're done stimulating a receptor, they are transported back into a cell to be broken down and recycled.

The chemical messengers of the ECS are the endocannabinoids. Other body systems employ chemical messengers: the endocrine system has hormones; the nervous system has neurotransmitters; the immune system has chemical messengers called cytokines; and the digestive system has gut peptides. Each chemical messenger has a different name based on its location, but they are actually the same chemicals. Just like in real estate, it's all about location! Serotonin in your gut is called a gut peptide, but in the nervous system it's called a neurotransmitter. Endocannabinoids are mainly found in synapses, blood, and extracellular fluid.

Endocannabinoids are unique in that they are not made by a particular gland for use at a distant target, like in the endocrine system. Each and every cell that has synthesizing enzymes has the ability to make them, on demand, when needed. Yes, each and every one of your cells has the ability to maintain its own balance when it deems it necessary.

We've recognized seven endocannabinoids: arachidonoyl ethanolamide (AEA), 2-arachidonoylglycerol (2AG), noladin ether, virodhamine, N-arachidonoyl dopamine, palmitoylethanolamide (PEA), and oleoylethanolamide (OEA). AEA and 2AG are the two best known and most studied, and they are the endocannabinoids we will focus our discussion on. Research is just beginning on the others.

Arachidonoyl Ethanolamide Anandamide (AEA)

AEA, the "bliss molecule," was the first endocannabinoid discovered. It is found in the brain, spleen, heart, skin, connective tissue, bone, and reproductive organs. It is a partial agonist of the CB1 receptor when found alone and binds with high affinity in the presence of lipoxin A4. AEA also binds to CB2, GRP55, TRPV1, and PPAR receptors. AEA is less abundant than 2AG in the brain, but concentrations quickly increase when needed. AEA is our very own THC. Yay!

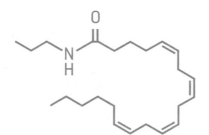

◄ AEA molecule

AEA SYNTHESIS AND BREAKDOWN

AEA is made from arachidonic acid, an essential fatty acid in the phospholipid bilayer of the cell, on demand, usually in response to a chemical cue. It is produced through a series of reactions with intermediate chemicals and corresponding enzymes, and is not stored for later use. Endocannabinoids cannot move in or out of the cell on their own; they require transport molecules to move them across cell membranes. AEA is shuttled around by fatty acid–binding protein (FABP). The enzyme fatty acid amide hydrolase (FAAH) breaks down AEA into arachidonic acid and ethanolamine. If FAAH is suppressed or there is too much AEA for FAAH to breakdown, AEA can remain available for further receptor binding or follow a different pathway and be broken down by cyclooxygenase-2 (COX-2) and subsequent enzymes into prostaglandin E2, an inflammatory prostaglandin. CBD is able to increase AEA levels by inhibiting the enzyme FAAH.

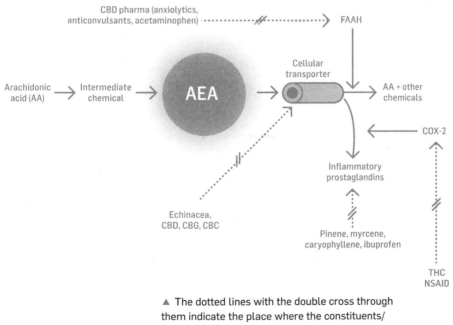

▲ The dotted lines with the double cross through them indicate the place where the constituents/herbs interact and *inhibit* the reaction.

2-Arachidonoylglycerol (2AG)

The endocannabinoid 2AG is a thousand times more abundant than AEA in the brain and a true agonist for the CB1 receptor; it binds with moderate to low affinity. It binds to the same receptors as AEA: CB1, CB2, GRP55, TRPV1, and PPAR.

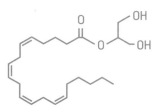

2AG SYNTHESIS AND BREAKDOWN

Production of 2AG also begins at the phospholipid bilayer in cell membranes. Through a series of reactions, it is derived from the precursor molecule diacylglycerol (DAG) and leaves the cell via its transport molecule, FABP, to bind to nearby receptors. It is broken down into arachidonate glycerol by the enzyme monoacylglycerol lipase (MAGL). It can also be broken down by COX-2, eventually producing PGE2, an inflammatory cytokine.

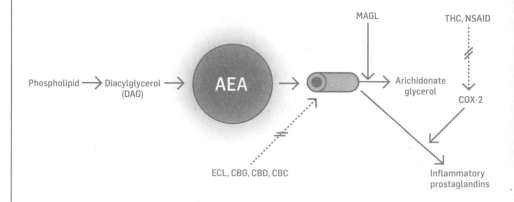

The Big Picture

Understanding these biochemical pathways will help us to understand how the multitude of interventions work. For example, ibuprofen and CBD both inhibit FAAH. This leads to less AEA being broken down, more AEA available to bind to receptors, and a decrease in inflammatory cytokine production. THC and CBD block FABP at multiple sites in the transport of AEA specifically, thus increasing AEA available to bind to receptors.

A Deeper Look at the Chemistry of the Endocannabinoid System

For those of you who want to go deeper into the chemistry of the endocannabinoid system, here is a more detailed look. If you're not interested in this level of detail, you can see a more general overview of endocannabinoids on page 89.

Endocannabinoid "Adjacents"

The endocannabinoids PEA and OEA are in the family of chemicals called N-acylethanolamines. They are made from fatty acids. They bind to receptors and are degraded by enzymes. While not considered true endocannabinoids — I call them "adjacents" because they don't bind to specific cannabinoid receptors — they do bind to some noncannabinoid receptors and function much like endocannabinoids.

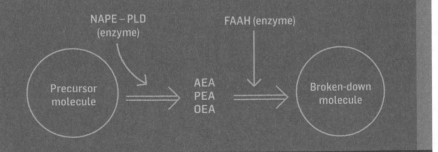

SUBSTRATE COMPETITION

Substances in the body compete for resources. OEA, PEA, and AEA all use the same starter molecules and enzymes for synthesis and breakdown, so quantities of one will affect numbers of the others. AEA, OEA, and PEA are all made using the synthetic enzyme NAPE-PLD. Making more OEA, for example, decreases the amount of synthesizing enzyme available to make AEA, decreasing AEA production and vice versa.

NAPE – PLD (enzyme)   FAAH (enzyme)

Precursor molecule → AEA PEA OEA → Broken-down molecule

Palmitoylethanolamide (PEA)

PEA binds to the PPAR, GRP55, and TRPV1 receptors; its effects are anti-inflammatory, antinociceptive, neuroprotective, and anticonvulsant.

Oleoylethanolamide (OEA)

OEA binds to the PPAR and TRPV1 receptors to stimulate fat break-down and regulate feeding.

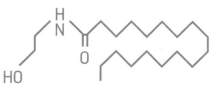

Chemical Functions of Endocannabinoids

In general, endocannabinoids inhibit the release of inflammatory chemicals and the release of neurotransmitters. AEA and 2AG function in similar ways when they bind to receptors. The major differences between them are where they work in the body and how strongly they bind to receptors. If we understand how our endocannabinoids work biochemically, we can apply that understanding to the cannabinoids of cannabis because they bind to the same receptors.

Endocannabinoids as Neurotransmitters

Information travels through the nerves of the nervous system via electrical currents called nerve impulses. Nerves are made of individual cells called neurons. Billions of neurons connect within the brain and spinal cord, but they don't quite touch each other. The tiny gap between neurons is called the synapse. Electric currents can travel through nerve fibers but cannot jump across synapses. That's where neurotransmitters come in. Neurotransmitters are chemical messengers released by an excited, "presynaptic" neuron carrying an electric current; these chemicals cross the synapse to bind to receptors on the "postsynaptic" neuron. This can either excite the postsynaptic neuron or inhibit it. The action of all endocannabinoids and cannabinoids happens right here at synapses. Now let's look at endocannabinoids as neurotransmitters.

In the nervous system both AEA and 2AG are made on demand at the postsynaptic neuron. When they are released from the postsynaptic neuron they travel back to the presynaptic neuron and bind to CB1 receptors on its surface. This prevents neurotransmitters stored in the presynaptic neuron from being released.

NEUROTRANSMITTERS AND THEIR EFFECTS

Acetylcholine: inhibitory and excitatory

Adenosine: inhibitory

Dopamine: emotional response, reward behavior, pleasure, skeletal muscle tone, movement

GABA: inhibitory

Glutamate: excitatory

Glycine: inhibitory

Norepinephrine: excitatory for awakening from deep sleep, dreaming, and mood

Serotonin: excitatory for sensory perception, temperature regulation, mood, appetite, and induction of sleep

Substance P: excitatory for pain perception

Endocannabinoids or cannabinoids binding to presynaptic CB1 receptors can be inhibitory or excitatory depending on which neurotransmitter they are preventing the release of. When they prevent an excitatory neurotransmitter such as glutamate from being released, the effect is inhibition. This is a protective mechanism that prevents the neuron from dying from overfiring (excitotoxicity).

When endocannabinoids or cannabinoids prevent the release of inhibitory molecules, GABA for example, the net effect is stimulation or excitation.

Along with binding to cannabinoid receptors, endocannabinoids can also bind to noncannabinoid receptors and modulate their shape to either increase or decrease their affinity for a particular ligand. For example, AEA is a positive allosteric modulator (enhance) of the glycine receptor and enhances its binding and activation. Glycine, when binding to its receptor, inhibits neuron activation. AEA enhances glycine's binding and thus its inhibitory function.

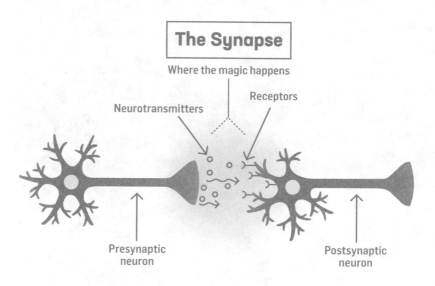

The Synapse

Where the magic happens

Neurotransmitters

Receptors

Presynaptic
neuron

Postsynaptic
neuron

▲ Cannabinoids and endocannabinoids play the vital role of chemical messengers; they can jump the tiny gap between nerves (called the synapse) and allow information to travel in the nervous system.

Endocannabinoids as Chemical Messengers

Endocannabinoids made by cells outside the nervous system act as chemical messengers, traveling to other cells and binding to their receptors to have an effect. (Hormones do this, too.) The net effect is decreased production of inflammatory cytokines (chemicals made by and/or for immune cells) and decreased migration of immune cells to inflamed areas. (Immune cells make inflammatory chemicals, which results in more inflammation.)

GENERAL OVERVIEW OF ENDOCANNABINOIDS

* Made on demand from precursor molecules in cell membranes and immediately released
* Broken down by enzymes: AEA by FAAH, 2AG by MAGL
* Retrograde (travel backward across the synapse) messengers in the nervous system
* Usually inhibitory
* Can be excitatory
* Chemical mediators of inflammation throughout the body

Receptors

To date, we know of 15 types of receptors in the ECS system. Five are "typical" cannabinoid receptors that work directly with endocannabinoids. The 10 others are not specific to the endocannabinoid system but are affected by endocannabinoids. Endocannabinoids affect these "atypical" receptors by changing the way the receptor interacts with its own specific chemical messenger.

The five typical cannabinoid receptors are CB1, CB2, GPR18, GPR19, and GPR55. The atypical receptors are GABA, serotonin (5-HT1A, 2A, 3), dopamine, adenosine, acetylcholine, glycine, glutamate, PPAR, TRP, and opioid.

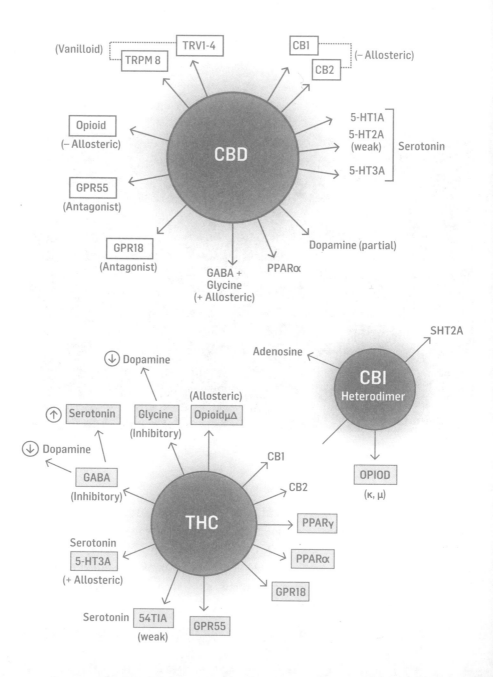

THE FLUID MOSAIC
CELL MEMBRANE

The membrane surrounding every single one of your cells is best described as a fluid mosaic. The membrane is a double layer of phospholipids that look like lollipops with their heads facing away from the cell nucleus and the stick tails facing inward. Those "sticks" are actually made of essential fatty acids (arachidonic acid and eicosapentaenoic acid, for example) that are the precursors to making endocannabinoids and inflammatory or anti-inflammatory cytokines. Interspersed within the cell membrane are receptors, transporters, and channels for activating cellular functions and transporting chemicals in and out of the cell. While we might draw diagrams of the cell that look static, the entire cell membrane is fluid, moving all the time. The "lollipops" are constantly moving around the perimeter of the cell or flipping to face the inside or outside of the cell. Receptors also move around the membrane. They can migrate to areas where there are more chemical messengers present or move into or outside the cell. When a receptor moves into a cell, it deactivates because it can no longer interact with a chemical messenger. Cells move receptors inside their membranes to regulate their sensitivity to a particular chemical messenger. It is miraculous how much control a single cell has over its own life!

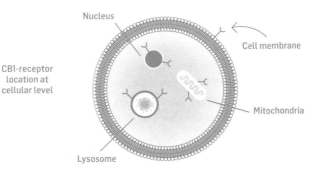

Nucleus

Cell membrane

CB1-receptor location at cellular level

Mitochondria

Lysosome

Lock and Key

In chemistry class, you may have been taught the "lock and key" model for how a chemical messenger binds to a receptor: a specific shape of molecule binds to a receptor specifically shaped to receive it. While this basic premise holds true, the model needs an upgrade. Both the ligand molecules and the receptors are dynamic, vibrating, wiggling, shimmying (dare I say singing?), and changing shape. They are not static; they are constantly moving!

If you'd like to expand your thinking around receptors and binding, there is evidence that it is not even necessary for a chemical messenger and receptor to physically bind. When the vibrating ligand comes close enough to the proper receptor, they entrain to the same vibrations, which activates the receptor. Right now, science has no way to measure receptor entraining because of the limitations of measuring equipment. One day Western science will be able to explain the healing that occurs when we know the specific healthy vibration for each receptor, cell, or tissue and can create that healing vibration and move it close enough to the "unhealthy" tissue so it can entrain the entire body to the song of health. Some healers are already doing this, in fact, and have been doing it for thousands of years in practices like qigong and hands-on healing.

Until science can explain this phenomenon, let's return to what is understood now. Receptors have more than one place for molecules to bind to. The primary active site is called the orthosteric site; other sites are called allosteric binding sites. Molecules binding to allosteric sites change the receptor's shape, which can enhance, decrease, or do nothing to the ability of chemical messengers to bind to the orthosteric site.

For example, the shape of the CB1 receptor is perfect for 2AG to bind to. That is why 2AG is an agonist with higher affinity to the CB1 receptor than AEA. AEA, on the other hand, is a partial agonist. It doesn't bind to the CB1 receptor with as much affinity as 2AG because AEA's shape is not as close a fit. Lipoxin A4, an anti-inflammatory metabolite of arachidonic acid, binds to a different allosteric site on the receptor that changes the receptor's shape to allow a better fit for AEA. Lipoxin A4 is, therefore, an allosteric enhancer of AEA. CBD binds to an allosteric site on CB1 and changes its shape so THC does not bind as well, so CBD is a negative allosteric modulator for THC. This will translate to some of the modulatory effects of CBD on THC's actions in the body.

Binding Terms

All chemicals are not created equal in how well they bind to receptors. There is a range of binding strengths; some chemical messengers bind with vigor and high affinity, while others barely hold on. Think of binding as hugging. Some huggers give you a full-body bear hug, while others barely embrace you. An agonist is a chemical that binds to a receptor and activates the receptor to produce a biological response. Think of an agonist as an "actor."

Full agonist: a chemical that binds to a receptor to produce a biological response. For example, synthetic THC is a full agonist of CB1 and binds to the receptor so tightly it can cause more intense effects than other partial agonists that don't bind to the receptor (AEA or THC) as strongly.

Partial agonist: a chemical that binds to receptors but only causes a partial activity. AEA, 2AG, and THC are partial agonists of CB1.

Neutral agonist: a chemical that binds to a receptor and causes no reaction. Neutral agonists work by displacing agonists (or blocking them from binding), causing a decrease of function of the receptor being bound by the agonist.

Inverse agonist: an inverse agonist binds to receptors and causes the opposite reaction as an agonist.

Positive allosteric modulator: a chemical that binds to secondary sites on a receptor and increases agonist signaling by either increasing agonist binding or increasing receptor sensitivity.

Negative allosteric modulator: a chemical that binds to secondary sites on a receptor and decreases agonist signaling by either decreasing agonist binding or increasing receptor sensitivity.

Antagonist: a chemical that blocks the action of the agonist.

Practical Applications of Understanding Receptors and Their Chemical Messengers

The binding of chemicals to cannabinoid receptors is not linear or binary. If you fill the receptors with chemical messengers, the effects are different than if you bind only a few receptors. Adding to the complexity is that the cannabinoid receptors regulate the release of other neurotransmitters and the fact that CB1 receptors can form heterodimers (pair up with) with 10 other receptors to have inhibitory or excitatory effects. This system is vastly more complex than the existing literature would lead us to believe, and the oversimplification of working with isolated constituents could be potentially dangerous given our binary understanding of a mycelial-like multifaceted design (keep this in mind when we discuss the drug rimonabant).

Cannabinoid Receptors

Cannabinoid receptors are found in all vertebrates, as well as many invertebrates and plants, but they are not found in insects (bugs cannot get high!). They are among the oldest type of receptors, dating back 600 million years. They are part of a general class of receptors called G-coupled receptors. All G-coupled receptors span the cell membrane, and the part of the receptor spanning the membrane is made of the same molecules. The difference between the various G-coupled receptors in the body (including CB1, CB2, GABA, dopamine, and serotonin) are the parts that stick out of the membrane to bind with chemical messengers. These parts are like catcher's mitts that will activate only with the one specifically shaped chemical messenger they can catch. Once a chemical messenger binds to a receptor, a cascade of reactions occurs within the cell to bring about an intended action (the functions we've been discussing).

Receptor Binding

The receptor is where the action is. All the functions of the ECS occur because of endocannabinoids binding to receptors. The effects of cannabis occur here, too. There are 15 different kinds of receptors within the ECS, and these receptors are found in virtually every tissue of the body, which makes the ECS a major regulatory system underlying the nervous, endocrine, immune, digestive, reproductive, and circulatory systems.

Every cell with receptors is a potential target. Chemical messengers that bind with a strong affinity have a strong effect; those with a weak affinity have a lesser effect. Furthermore, the strength of the bond can be affected by allosteric binding molecules.

We can apply what we know about our endocannabinoid system to the workings of cannabis and her chemical constituents. Cannabis communicates to us when her cannabinoids bind to our ECS receptors. The structural similarity of cannabis's constituents is so close to our own molecules that they bind to our receptors and produce effects like our own *endo*cannabinoids.

Cannabinoid 1 Receptor (CB1)

Named CB1 because it was the first receptor discovered in the ECS, CB1 receptors in the brain are found in the regions of the amygdala, hippocampus, cerebellum, cerebral cortex, olfactory bulb, and basal ganglia. No CB1 receptors exist in the brain stem, which controls breathing and heart rate. This is why you cannot die from an overdose of cannabis. Opiates, on the other hand, do bind to receptors in the brain stem. When too many opiate molecules (overdose) bind the receptors here, your breathing and heart will stop.

CB1 receptors are also found in peripheral nerves, the thyroid gland, adipocytes, uterus, pituitary gland, hepatocytes, adrenal gland, reproductive organs, skeletal muscle, lungs, bone marrow, bone tissue, bladder, pancreas, oligodendrocytes, microglia, astrocytes, skin neurons, immune cells, and the gastrointestinal tract.

CB1 Functions

The CB1 receptor is the most abundant G-coupled protein receptor in the brain, 10 to 50 times more abundant than opiate and dopamine receptors. When bound, the receptor inhibits neurotransmitter release from the neuron. This regulates learning, memory, emotional response, body temperature, motor function, reward, and addiction. In the rest of the body it regulates appetite, metabolism and food intake, bone mass, and tumor cells of neuroglia and endothelium. It also reduces neuroinflammation and pain, and controls differentiation in the developing brain and neuronal survival and synaptic plasticity in the developed brain.

Binding to CB1

Each receptor has a specific chemical messenger that binds to it. It is important to understand which chemicals bind to which receptors because this will help us create effective medicine for particular conditions.

Partial agonist with higher affinity: 2AG, THC
Partial agonist with weaker affinity: AEA, CBC, CBG, CBN
Negative allosteric modulator: CBD
Inverse agonist: CBD
Heteromeric with: serotonin, dopamine, adenosine, opioid, orexin, and chemokine receptors

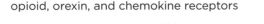

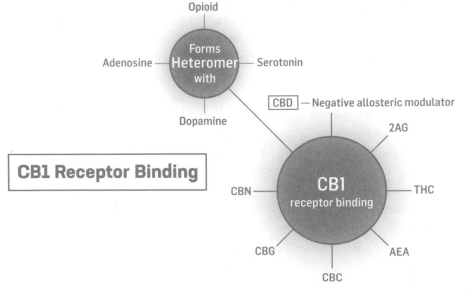

CB1 Receptor Binding

Location of CB1 Receptors

To understand the function of a receptor, we need to know where it is.

AT THE CELL MEMBRANE

Cannabinoid receptors exist on the surface of cells within the cell membrane.

Neurons. When CB1 receptors are bound by their ligands on neurons, they inhibit neurotransmitter release from the neuron. CB1 receptors are found in neurons that release GABA, glutamate, serotonin, dopamine, acetylcholine, norepinephrine, corticotropin-releasing factor (CRF), cholecystokinin, dynorphin, and substance P. We will discuss this in detail in the chapter on conditions (page 164).

DIMERIZATION

The concept of dimerization — the process of two receptors joining together — is not often discussed in introductory biology, but it's an important factor in the CB1 receptors' functioning. CB1 receptors can move within the cell membrane and link with other G-coupled protein receptors (all receptors can move within cell membranes). This linkage changes the receptors' affinity for chemical messengers, their signaling, and even their placement. If you recall how allosteric modulators did this (page 92), heteromers — two joined receptors — do the same thing but on a larger scale. To date, we have found that the CB1 receptor can dimerize with many different receptors, including those for dopamine, 5-HT2A serotonin, orexin, opioids, adenosine, and chemokines.

The science community's understanding of CB1 heteromers continues to unfold, but we know that this mechanism is involved with opiate tolerance, depression associated with pain, cancer cell metastasis, and the negative effects of cannabis on memory.

Immune Cells. Chemical messengers binding to CB1 receptors on immune cells usually decreases the production and release of inflammatory cytokines and chemokines that attract more immune cells. The net effect is a decrease in inflammation.

Endothelial Cells. Endothelial cells line structures like blood vessels. Chemical messengers binding to CB1 receptors causes an increase in nitric oxide (NO) production, which causes vasodilation (dilation of the blood vessels, which lowers blood pressure) and decreased platelet aggregation.

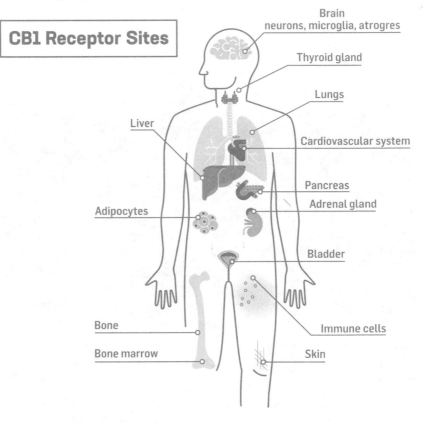

CB1 Receptor Sites

Brain
neurons, microglia, atrogres

Thyroid gland

Lungs

Liver

Cardiovascular system

Pancreas

Adrenal gland

Adipocytes

Bladder

Bone

Immune cells

Bone marrow

Skin

INTRACELLULAR CB1 RECEPTORS

In addition to cannabinoid receptors on the surface of cells, receptors have also been discovered within the cell on the mitochondria, lysosome, and the nucleus.

Mitochondria. When ligands bind to mitochondrial receptors, cellular respiration and cyclic AMP function decreases. Because mitochondria

LIFE WITHOUT CB1 RECEPTORS

Research animals whose CB1 receptors were permanently blocked became anxious and depressed and had increased mortality. They became regressive, could not enjoy pleasure, were afraid of the new, and couldn't unlearn fear. Studies show that individuals with phobias, chronic pain, and post-traumatic stress disorder (PTSD) have fewer CB1 receptors or their receptors have a weaker binding capacity. In a chemical messenger-receptor system, decreasing the number of functional receptors decreases the function of the whole system. RImonabant was a pharmaceutical CB1 blocker introduced for weight loss that was pulled from the market when users suffered an increase in depression and suicide.

use oxygen to produce energy, the cell manufactures less energy. Evidence suggests that this might contribute to memory loss due to neurons' decreased ability to manufacture energy in the mitochondria.

Lysosomes. CB1 receptors have also been found within the cell on the surface of the tiny organelle called the lysosome. Lysosomes carry degradation enzymes that help the cell break down bacteria or dysfunctional cell parts — think of them as internal cellular recycling centers. They also can cause the release of intracellular calcium from other organelles. Cells undergoing autophagy increase the use of lysosomes.

Nucleus. Ligands binding to CB1 receptors on the nucleus helps regulate cell proliferation, cell death (apoptosis), cell differentiation in a developing brain, and neuronal survival in a developed brain. In neurons, binding protects against nutrient deprivation and neuronal degeneration.

CB1 Chemistry

The CB1 receptor has constitutive activity, meaning it's active, just a little bit, all the time, even in the absence of an agonist. Blocking this low-level activity with an inverse agonist decreases important baseline functions, in this case GABA signaling (GABA is an inhibitory neurotransmitter linked to anxiety). CB1 signaling increases the amount of GABA available, decreasing anxiety. The pharmaceutical rimonabant was in clinical trials for weight loss as a CB1 inverse agonist, but the drug was pulled when participants in studies became suicidal and depressed. The low-level CB1 receptor activity is needed at all times and should not be blocked.

The CB1 receptors, like all G-coupled protein receptors, float within the cell membrane. The cell membrane is more fluid than solid, with receptors and individual fatty acids constantly moving around. The entire plasma membrane of some cells is renewed each hour through the fatty acids flipping in (endocytosis) and out (exocytosis) of the cell membrane: hundreds of millions of fatty acids and tens of thousands of receptors. Wow! When the CB1 receptor, or any receptor, is endocytosed, it is removed from active duty and is downregulated. When cells downregulate receptors, they become less sensitive to agonists. Up- or downregulating cell receptors is a primary way of regulating chemical messenger function at the cellular level. Not all CB1 receptors are downregulated equally throughout the brain. This uneven downregulation explains why tolerance grows for some effects of THC but not others.

Receptor-binding nerd fact: 50 percent of pharmaceuticals use the general class of G-coupled receptors. They don't use the cannabinoid receptor; they use the same type of receptor (G-coupled). The basic formation and building blocks of all the G-coupled receptors are the same. This could become a problem if someone is taking many pharmaceuticals. When the body needs to make more of one type of receptor (because someone is taking a pharmaceutical that binds it), it will pull resources from a limited pool. Since the cannabinoid receptors are made of the same materials (in a slightly different shape), your body may not have the resources it needs to make all the receptors it needs. Stealing parts from one receptor to make others is called "bullying" (for real). If one type of receptor is overactive (like when you're taking many pharmaceuticals), the cell can steal parts needed to make more of the overactive receptor from other receptors. If the cell steals receptor parts from your endocannabinoid (or any other) receptors, you will make fewer of those receptors and therefore tend toward a "hypo" situation, with fewer receptors available for endocannabinoids to bind to. We will see why this matters when we discuss ECS deficiency. It is important to look at the number of pharmaceuticals an individual is taking to discern if bullying of the cannabinoid receptors is occurring.

Cannabinoid 2 Receptor (CB2)

The CB2 receptor is found primarily in the immune system on natural killer cells, B cells, mast cells, macrophages, T cells, tonsils, spleen, thymus, and liver (Kupffer cells). They are also found in cardiovascular tissue (myocardium and endothelium), skin, reproductive tissue, bone, connective tissue, endocrine and exocrine pancreas, tumors, brain, and in the gastrointestinal tract. In the nervous system they are found on microglial cells. Like the CB1 receptors, CB2 receptors are inhibitory when they are bound to by their chemical messenger. CB2 receptors are not commonly found in the nervous system, but when nerve tissue is inflamed or injured, CB2 receptor number increases in neurons, microglia, and astrocytes.

CB2 Functions
Binding CB2 receptors protects against osteoporosis, atherosclerosis, pain (nociception), chronic liver disease, neurodegeneration, metabolic disorders, inflammation, neuroinflammation, and drug addiction.

Binding to CB2
Full agonists: AEA, 2AG, THC, caryophyllene (terpene)
Weak agonists: CBN, CBC, CBG
Inverse agonist: CBD

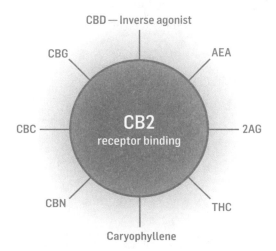

◀ When CB2 receptors are bound to by their chemical messengers, this binding helps protect against pain, atherosclerosis, neuroinflammation, drug addiction, and other disorders.

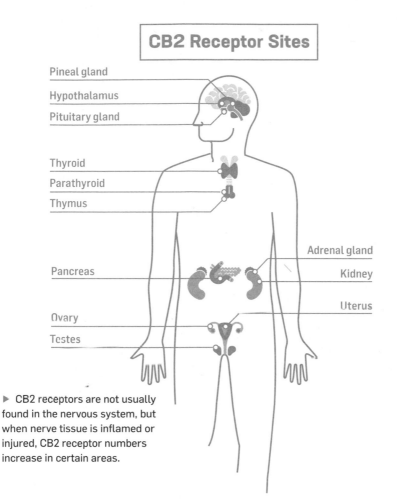

CB2 Receptor Sites

Pineal gland

Hypothalamus

Pituitary gland

Thyroid

Parathyroid

Thymus

Adrenal gland

Pancreas

Kidney

Uterus

Ovary

Testes

▶ CB2 receptors are not usually found in the nervous system, but when nerve tissue is inflamed or injured, CB2 receptor numbers increase in certain areas.

Atypical Cannabinoid Receptors

There are nine "atypical" cannabinoid receptors, so called because they were discovered before the endocannabinoid system and associated with other systems. They are either bound to directly at the orthosteric site (the active site where the chemical messengers bind) by endocannabinoids or cannabinoids or at allosteric sites. In a few instances endocannabinoids/cannabinoids affect transport of the natural chemical messenger of the receptor, which ultimately affects receptor function. These nine receptors add to the depth and complexity of the endocannabinoid system and the places cannabis interfaces with it.

GPR18 Receptor

This is called an "orphan" receptor because it was not understood that GPR18 is a G-protein receptor like CB1 and CB2, that interacts with the endocannabinoids. This receptor was "adopted" in 2006 when it was found that AEA binds to it weakly and N-arachidonoyl glycerine, a metabolite of AEA, binds to it strongly. We also know now that THC binds to the GPR18 receptor and CBD is an antagonist. The GPR18 receptor is found in the spinal cord, small intestine, immune cells, spleen, bone marrow, thymus, lungs, testes, and cerebellum. Its functions include blood pressure regulation, and it aids in immune function as a chemoattractant (binding of it attracts immune cells).

GPR55 Receptor

Another orphan, GPR55 receptors were "adopted" in 2007 as endo-cannabinoid receptors of AEA, 2AG, and THC. CBD antagonizes the receptor. GPR55 receptors are found in the central nervous system, adrenal glands, gastrointestinal tract, lungs, liver, uterus, bladder, and kidney. Binding decreases blood pressure and inflammation, blocks some pain, regulates energy intake and expenditure (it may play a role in obesity and diabetes mellitus), regulates bone cell function, and decreases neurodegeneration.

GPR19 Receptor

The last of the orphaned receptors, GPR19 was found in 2006 to bind to AEA and 2AG weakly; OEA is a full agonist. This receptor is limited to the pancreas and the gastrointestinal tract. It functions to decrease food intake, decrease body weight, and improve blood sugar.

Acetylcholine Receptors

Acetylcholine receptors can be either excitatory or inhibitory depending on their location. Two subtypes of the receptor exist: the nicotinic ion-gated channel and the muscarinic G-protein receptor. We will limit our discussion to the nicotinic receptor: this is where the cognitive effects of nicotine are triggered. Acetylcholine receptors are found in the brain (cerebral cortex, thalamus, and hippocampus), neuromuscular junctions, and autonomic nerve ganglia. Acetylcholine receptors affect memory, learning, and attention.

BINDING TO ACETYLCHOLINE RECEPTORS

Noncompetitive agonist: AEA

Negative allosteric modulator: CBD

Adenosine Receptors

Binding of adenosine receptors promotes rest in neurons of the brain and spinal cord by decreasing nerve transmission, so their binding has an inhibitory effect. Caffeine antagonizes this receptor, blocking adenosine from binding to it (and keeping neurons from getting the break they need!). Adenosine receptors are found in the region of the brain called the dorsal striatum, which is involved in motor activity, cognitive function, and mood.

Adenosine binding causes drowsiness and memory impairment, and it affects cognitive function, mood, and motor activity. (Think of all the things the wonder drug caffeine does for us as it blocks adenosine function — we are able to stay up longer, think faster, and write books!)

The neurotransmitter adenosine supports sleep, suppresses arousal, and is a vasodilator. When it is able to bind to its A2A receptor, it is anti-inflammatory. Increased adenosine is also responsible for the decreased anxiety and better sleep promoted by CBD and THC. They both inhibit adenosine's transporter and decrease the amount of adenosine being removed from the synapse, allowing it to carry out its functions.

Dopamine and Receptors

The ECS, through binding glutamate and GABA, affects dopamine levels rather than interacting with dopamine receptors directly. Dopamine is one of the more familiar neurotransmitters, associated with reward and mood. The endocannabinoid system regulates dopamine in a few different ways, and it is dopamine that mediates some of the effects of THC, such as the munchies, lowered cognition, psychosis, amotivation in some heavy users, and the reinforcement behavior of addiction. It is interesting to note that cannabis increases dopamine levels much less than the 10 to 30 percent jump in dopamine levels that cocaine and amphetamines cause. Dopamine receptors are found in the brain in the regions of the midbrain, pre-frontal cortex, nucleus accumbens, dorsal striatum, and basal ganglia. Their binding functions in mood, arousal, learning, motivation, and coordinating actions to attain a reward.

CB1 agonists (THC, 2AG, AEA) can either increase or decrease dopamine levels depending on which dopamine-releasing neurons they are connected to. CB1 agonists binding to glutamate receptors cause an increase in the release of dopamine, whereas CB1 agonists binding to GABA receptors cause a decrease in the release of dopamine. Dopamine neurons make endocannabinoids that feed back to glutamate receptors and GABA receptors to regulate the release of dopamine.

CBD is a partial agonist of the dopamine receptor and may be responsible for the effects of drowsiness, fatigue, diarrhea, and decreased appetite.

Gamma-Aminobutyric Acid (GABA) and Receptors

The neurotransmitter GABA is inhibitory like glycine but primar-ily functions in the brain, whereas glycine is the primary inhibitory neurotransmitter in the spinal cord. The inhibitory functions asso-ciated with GABA are sedation, decreased anxiety, and antiseizure properties. Higher levels of GABA receptors are found in the brain, spinal cord, and pancreas. Lower levels can be found in the intestines, stomach, fallopian tubes, uterus, ovaries, testes, kidney, bladder, lungs, liver, and immune cells.

GABA receptor activation is the major inhibitory mechanism of the central nervous system. Think of it as our neuronal off switch. It regulates pain-signal transduction and causes sedation. Alcohol depresses the nervous system via GABA activation. People with GABA deficits are more susceptible to the psychomimetic effects of THC. Withdrawal from alcohol and benzodiazepines is caused by decreased GABA in the system; when a person consumes external GABA receptor stimulators, the body makes less GABA of its own to maintain balance.

BINDING TO GABA RECEPTORS

Agonists: CBD and CBDA (less sedation than pharma while maintaining anxiolytic effects), barbiturates, benzodiazepines, alcohol, antiseizure meds

Positive allosteric modulators: CBD and CBDA

THC increases GABA via decreasing reuptake.

Glutamate Receptors

Glutamate is the most abundant excitatory neurotransmitter found in the brain. It is involved in most aspects of normal brain function including cognition, memory, learning, and motor activity. Too much glutamate can cause damaging excitotoxicity; interestingly, too little glutamate causes difficulty concentrating and mental exhaustion.

The main receptor for glutamate is the NMDA receptor. When a CB1 receptor is bound by its agonists, on NMDA-containing neurons, a decrease in glutamate binding occurs. Further, binding of CB1 can cause a decrease in NMDA receptors on neurons as well. All this aids in preventing glutamate-induced excitotoxicity.

Glycine Receptors

The glycine receptors are inhibitory ligand-gated ion channels. Glycine receptor channels can be found in the brain (hippocampus, amygdala, cerebral cortex) and spinal cord. Binding the glycine receptor has the effects of neuroprotection, anti-inflammation, decrease in pain signal transmission, and decrease in release of dopamine.

BINDING TO GLYCINE RECEPTORS

Agonists: glycine, AEA, 2AG, OEA, THC (low dose), and CBD
Positive allosteric modulators: THC and CBD

Opioid Receptors (OR)

The binding of ligands to the opioid receptors causes a decrease in pain transmission within the peripheral and central nervous system. There are four receptor subtypes — kappa, delta, gamma, and mu — and they are typically found on presynaptic neurons.

Endocannabinoids regulate pain through a variety of ways (addressed in the chapter on conditions, page 164); opioid receptors are one such pathway. Opioid receptors can be found in the brain, spinal cord, peripheral nerves, and digestive tract. In general, binding of opioid receptors decreases pain signaling by inhibiting the release of substance P and glutamate.

Delta-OR

Binding of the delta-OR is responsible for analgesic, antidepressant, and physical dependence effects of opiates.

BINDING TO DELTA OR

Agonists: endorphins, enkephalins, dynorphins

Negative allosteric modulator: CBD/THC (possible mechanism of opiate sparing, prevention of tolerance, and help with opiate withdrawal)

Heterodimer with CB1: inhibit each other's individual binding functions. Low-dose CB1 agonist increases activity of delta-OR.

Kappa OR

Binding of the kappa OR has analgesic and anticonvulsant effects in the body; it also affects depression, hallucinations, diuresis, neuropathic pain, and sedation.

BINDING TO KAPPA OR

Agonists: endorphins, enkephalins, dynorphins

Heterodimer with CB1

Mu OR

Binding of the mu OR is responsible for analgesia, physical dependence, respiratory depression, euphoria, and decreased gastrointestinal motility.

BINDING TO MU OR

Agonists: endorphins, enkephalins, dynorphins, opiates (note, these are the only opiate receptor that opiates bind to)

Negative allosteric modulator: CBD/THC

Peroxisome Proliferator-Activated Receptors (PPAR)

We now move away from the cell-surface receptors and into the nuclear receptors. PPAR receptors are further divided into two subcategories, PPAR alpha and PPAR gamma. All PPAR receptors are found within a cell's nucleus. PPAR alpha receptors are found in the liver, kidney, heart, skeletal muscle, and adipose tissue. PPAR gamma receptors are found in the heart, muscles, colon, kidney, pancreas, and spleen.

These receptors affect expression by the DNA of regulatory proteins of the cell. Specifically, binding of the PPAR receptors by their agonists regulates lipid metabolism, insulin sensitivity, glucose metabolism, inflammation, pain, cell proliferation, and hepatic enzyme expression.

BINDING TO PPAR

PPAR alpha weak agonists: fatty acids, eicosanoids, 2AG, AEA, PEA, OEA, THC, CBD, CBC, CBG

PPAR gamma weak agonists: AEA, 2AG, OEA, PEA, THC

Serotonin Receptors (5-HT)

Binding of serotonin receptors regulates blood pressure, heart rate, mood, learning, memory, sleep, body temperature, appetite, nausea, vomiting, cerebral blood flow, and acute responses to stress. Binding the receptor is anxiolytic, panicolytic, and antidepressive.

Serotonin receptors are found throughout the body. All of the various functions of serotonin are carried out by the binding of seven different subtypes of the serotonin receptor. The binding of the serotonin receptors by its agonists can either be inhibitory or excitatory for the release of the following neurotransmitters: glutamate, GABA, dopamine, epinephrine and norepinephrine, acetylcholine, oxytocin, prolactin, vasopressin, cortisol, and substance P.

The serotonin (5-hydroxytryptamine or 5-HT) receptors are divided into subclasses of 5-HT1 (A-F), 5-HT2, and 5-HT3. The different types of 5-HT receptors are categorized by location and function.

Pharmaceuticals that target the serotonin receptor are antidepressants (serotonin reuptake inhibitors), antipsychotics (aripiprazole), anorectics, antiemetics, antimigraines, and hallucinogens. CB1 receptor activation at serotonin synapses, GABA, and glutamate receptors increases the release of serotonin.

Serotonin Receptors

Receptor type	Where found	What binds to it	Functions
5-HT1A	blood vessels, brain/spinal cord	agonist: CBD, CBDA; THC partial; antagonist CBG	antidepressant, anxiolytic, mood, sleep, vasoconstriction
5-HT2A	blood vessels, CNS, GI, platelets, PNS, smooth muscle	CBD (weak), LSD, mescaline, DMT, psilocybin, CB1 heterodimer	addiction, anxiety, appetite, cognition, learning, memory, mood, sleep, vasoconstriction
5-HT3A	CNS, GI, PNS	allosteric modulator CBD/THC, AEA antagonist	anxiety, addiction, emesis, GI motility, learning, memory, nausea

5-HT1A

The 5-HT1A serotonin receptor is found in the brain (cerebral cortex, hippocampus, amygdala, septum, raphe nucleus). Binding of this receptor lowers blood pressure, heart rate, and body temperature. It is antiemetic, analgesic, and antinausea; it increases cerebral blood flow and attenuates acute response to stress; and it is anxiolytic and panicolytic and an antidepressant.

BINDING TO 5-HT1A

Agonists: CBDA (100 times stronger than CBD), CBD; pharma: buspirone, SSRI, MDMA
Partial agonist: THC
Antagonist: CBG

5-HT2A

The 5-HT2A serotonin receptor is also found in the brain's cerebral cortex, hippocampus, amygdala, septum, and raphe nucleus. Binding of this receptor affects emotions, learning and memory, and pain.

BINDING TO 5-HT2A

Agonists: entheogens (psychedelics): mescaline, psilocybin, DMT, LSD; pharma: antidepressants and antipsychotics

Weak agonist: CBD

Heterodimer: low-dose THC lowers anxiety; high-dose THC affects memory

5-HT3A

The 5-HT3A receptor is found in the brain, spinal cord, peripheral nerves, and gastrointestinal tract. Binding the receptor affects pain, mood, nausea, and emesis.

BINDING TO 5-HT3A

Antagonist: AEA

Negative allosteric modulators: CBD, THC

Transient Receptor Potential (TRP) Cation Channels

The TRP channels are ion channels found on the cell membranes. There are roughly 30 different types of TRP receptors in the body, and depending on the type are found in the central nervous system, on sensory neurons, immune cells (macrophages, dendritic cells, Langerhans cells), endothelium, epithelium, epidermis, hair follicles, keratinocytes (skin cells), perivascular tissue, intestines, kidney, placenta, spleen, lung, and smooth muscle. The vanilloid TRPV group is colocalized with CB1 and CB2 receptors in sensory neurons in the brain and skin; we will focus on this group.

TRPV (VANILLOID) RECEPTOR

The vanilloid receptors are a class of cell membrane receptors that signal burning pain, cause vasodilation of blood vessels in the area, and lower the temperature of the body.

When the vanilloid receptor is activated you feel a burning pain. Remember the last time you had too much wasabi or cayenne? That's the burning pain. Notice the redness on your skin when you get a little on it? That's vasodilation. Dilation of the blood vessel brings in blood. The sensation of pain is protective. It alerts you that something different needs to be done: "Move your hand away from the hot water!" or "Stop rubbing cayenne in your eye!"

The vanilloid receptor can become sensitized/activated by acidic conditions like inflammation and ischemia. This signals pain, helping you protect the area. Inflammatory chemicals such as bradykinin, serotonin, histamine, and prostaglandins also sensitize the receptor. Other agonist/sensitizers of the receptors are heat (above 109°F, or 43°C), menthol, piperine (black pepper), zingerone (ginger), and iso-thiocyanate (wasabi and mustard).

Vanilloid receptors can also become desensitized/deactivated. People employ this biochemistry when they use a cayenne or menthol cream or salve on painful joints in the body. The continual stimulation by capsaicin in cayenne or menthol quiets the pain signal moving to the brain. Continual stimulation of the vanilloid receptor by capsaicin, AEA, OEA, or CBD all desensitize the receptor and decrease the perception of pain. CBD can also be used internally for pain management at the TRPV1 receptor. An added benefit of cannabis is that it doesn't cause the burning sensation topically that cayenne does.

Another way to manage pain is to decrease or downregulate the number of receptors. If there are fewer receptors that signal pain, there will be fewer pain signals, which results in less pain. In astrocytes and microglia, chemicals called specialized pro-resolving mediators act on other receptors to decrease the total number of vanilloid receptors that signal pain.

BINDING TO TRP

Agonists: AEA, 2AG, THC, THCV, CBD, CBN, CBC, CBG, PEA, and OEA all desensitize at least one TRP receptor. CBD is the only constituent in cannabis that desensitizes all TRP receptors. Other agonists include heat above 109°F or (43°C), low pH, cayenne (capsaicin), mint (menthol), pepper (piperine), ginger (zingerone), and wasabi (isothiocyanate).

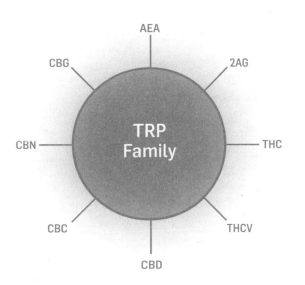

Vanilloid receptors	What binds	Where found	Functions
TRPV1	PEA, OEA, 2AG, AEA, CBD, DRG, PNG, CBC, THCV, CBN, CBG (antagonist)	dorsal root ganglia (DRG), PNG, trigeminal ganglia (TG), testes, bladder, skin, pancreas	Pain, nociception, temperature sensation
TRPV2	THC, CBD, CBC, CBG (antagonist)	DRG, brain, spleen, GI, mast cells; smooth, cardiac, and skeletal muscle cells	Nociception, analgesia, temperature perception, antiproliferation
TRPV3	THC, CBD, CBC, THCV, CBDV, CBG	DRG, TG, CNS, skin, tongue, testes, hair follicles	Nociception, analgesia, temperature perception, antiproliferation
TRPVA1	CBD, CBN, CBG, CBC, THC	DRG, TG, hair cells, fibroblasts, ovaries, spleen, testes, GI	Cold sensation, menthol, pathophysiological cold pain, inflammatory and nociceptive pain

How to Support the ECS

There are a number of ways to support the endocannabinoid system. Much like with the immune system, we can nurture a healthy ECS through diet and lifestyle habits.

Consume Omega-3 Fatty Acids

Omega-3 fatty acids are beneficial in three ways. They increase CB1 and CB2 receptors, they increase the endocannabinoid synthetic enzymes necessary for making 2AG and AEA, and they are responsible for proper signaling of the ECS. Docosahexaenoic acid (DHA) and eicosapentaenoic acid (EPA), both omega-3 fatty acids, are precursor molecules used by the body to make AEA. By consuming omega-3 fatty acids, we support the ECS in three ways: we increase the number of ligand building blocks, increase synthetic enzymes, and increase receptor numbers! Good sources of omega-3 fatty acids are free-range chicken, eggs, and cold-water fish such as cod and salmon.

It is important to note that in the case of obese individuals, excess omega-3 fatty acids can lower endocannabinoids, especially in dysregulated adipose and liver tissue. Non-obese people tend not to experience a decrease in endocannabinoids with omega-3 supplementation.

Downregulation of CB1 receptors and excess ECS signaling can occur with an excess of omega-6 fatty acids (arachidonic acid is an omega-6 fatty acid). The average American consumes 20 to 30 times more omega-6 fatty acids then omega-3s. This imbalance causes inflammation and pain. One surefire way to reduce omega-6 consumption is to stop eating factory-farmed animals and their eggs and milk. Animals raised in unnatural, unhealthy ways, eating foods they don't normally eat, like corn, make more omega-6 fatty acids. Animals raised and fed naturally (free ranging and grass grazing) make more omega-3 fatty acids.

Decrease Sugar and Unhealthy Fats

A diet high in trans fats and sugar leads to obesity and dysregulation of the ECS. It does so by causing the body to overproduce endocannabinoids, their synthesizing enzymes, and CB1 receptors in visceral adipose, liver, pancreas, and skeletal muscle. (Such a diet also causes diabetes mellitus, metabolic syndrome, atherosclerosis, and heart disease.) This increase in endocannabinoids causes continual CB1 activation, which results in an increase in lipogenesis, decreased insulin sensitivity, and blockage of glucose and fatty acid oxidation, which leads to glucose intolerance. The ECS is responsible for stimulating hunger; when the ECS is dysregulated, it prompts the body to eat more. Maintaining healthy body weight helps regulate the ECS.

Maintain a Healthy Gut

A healthy population of gut bacteria is essential. The bacteria in your gut modulate CB1 and CB2 expression and upregulate CB2 receptors. The best first step toward gut health is to get a good quality, one-month supply of probiotic supplement from the refrigerated section of your local health food store. After a month, continue to keep the gut healthy by eating fermented foods and whole plants to give the beneficial bacteria good food to eat. An added benefit of healthy gut flora is an increased expression of CB1 and mu opioid receptors in intestinal epithelium, making us more responsive to endocannabinoids and endorphins. Better responsiveness translates to feeling happier and safer. It makes sense that the bacteria in our bodies would want their host (us) to be happy and healthy so they can continue living the good life in and on us.

Consume Plant Helpers of CB1 and CB2

In addition to cannabis, there are many herbs and plants that manufacture chemicals that bind to cannabinoid receptors; they act as agonists to CB1 and CB2. CB1 agonists include copal, absinthe, salvia divinorum, camellia sinensis, kava, echinacea, legumes, club moss, algae, liverwort, and helichrysum, as well as some fungi. CB2 agonists include copal, echinacea, and rue.

Avoid Chemicals That Alter ECS Function

The insecticide pyrethrum, used in conventional agriculture, antagonizes the CB1 receptor. Phenyl phthalate, an additive used to mold plastics, antagonizes the CB1 receptor; it is also an endocrine disruptor and carcinogen. Simply put, try to eat organic and stop using plastic.

Manage Stress

Stress per se has gotten a bad rap. Our stress response is vital and necessary for health and learning. Our inability to manage stress is the problem — and can be a cause of serious disease. Acute stress (and subsequent increase in glucocorticoids and cortisol levels) enhances the ECS, while chronic stress and increased cortisol levels downregulate the ECS. Stress hormones decrease both AEA and 2AG in the corticolimbic circuit, the circuit that helps to regulate the amygdala (and our ability to modulate our emotions and stress response).

Move Your Body

Exercise keeps us healthy, helps us manage stress, is the number one intervention for mild depression, and increases ECS signaling by increasing serum AEA and CB1 receptor expression. The experience of "runner's high" is a combination of increased endorphins (our body's own opiates) and endocannabinoids. How do we know this? Researchers have chemically blocked endocannabinoids and endorphins both alone and together and found that both are responsible for the good feelings we receive as a reward for exercise.

Enjoy a Massage

AEA levels increased 168 percent in healthy individuals (asymptomatic for any disease state) when they received a general relaxation massage.

Be Aware of Pharmaceutical Levels

Anxiolytics increase AEA by inhibiting FAAH. Antidepressants, antipsychotics, and anticonvulsants increase CB1 receptors, which results in an increase in ECS tone and may be responsible for weight gain associated with these medications. Acute opiate use increases ECS function by increasing the synthesis of CB1 receptors. Chronic opiate use downregulates the ECS. Look for a larger discussion of cannabis and opiates in the chronic pain entry in the chapter on conditions (page 164).

Limit Alcohol

Acute alcohol use increases ECS signaling. Chronic or binge drinking downregulates the CB1 receptors. There is conflicting evidence of it both increasing and decreasing AEA and 2AG levels.

Limit Caffeine

The brain stimulation that comes with consuming caffeine is a result of caffeine blocking the adenosine receptor. Adenosine acts as a brake for overstimulated neurons. By blocking adenosine from binding to its receptor, caffeine prevents neurons from resting. Adenosine also inhibits CB1 receptors, allowing more dopamine and glutamate to be released — more happy feelings when we drink coffee! Although the ECS may be stimulated, it does not protect the adrenal glands from the detrimental effects of chronic caffeine overconsumption.

Endocannabinoid Deficiency Syndrome (ECDS)

The ECS may become dysregulated — meaning too many or too few endocannabinoids are being made by the body — following long-term perturbations. Suboptimal functioning of the ECS, termed endocannabinoid deficiency syndrome (ECDS), can be a factor in many diseases, including migraine headaches, irritable bowel syndrome, fibromyalgia, depression, anxiety, multiple sclerosis, Huntington's disease, chronic motion sickness, anorexia, schizophrenia, Parkinson's disease, failure to thrive, and PTSD. (The list is sure to keep growing as we understand more about the ECS.) Because so many people are affected by these conditions, it is useful to know how to improve and increase functioning of the ECS.

In general, for the ECS to work properly, the body must produce sufficient chemical messenger molecules and their receptors and remove the chemical messengers from circulation. When the body makes too much of any chemical messenger or too many receptors, the body is said to be in a hyper or overproductive disease state. The body retains some regulatory function by increasing or decreasing receptor numbers in response to increased or decreased chemical messengers, but the system can't completely compensate. Too little messenger or too few receptors causes underproductive (hypo) disease states. For example, in thyroid disease, too little thyroid hormone results in hypothyroidism and all the symptoms associated with it.

Because the endocannabinoid system underlies the autonomic nervous system, its dysfunction can be related to many disease states, especially when we consider the role inflammation plays in chronic disease and the role endocannabinoids play in regulating inflammation.

Aiding Endocannabinoid Deficiency Syndrome

All the steps we discussed for how to support the endocannabinoid system previously would be employed in ECDS. The basic approaches for ameliorating ECDS are much like our efforts to support any other receptor-chemical messenger system.

INCREASE CHEMICAL MESSENGER SYNTHESIS

Increasing chemical messenger synthesis can be accomplished by supplying the nutritional building blocks the body needs to make them. In the case of the endocannabinoids it would be the omega-3 fatty acids. This ensures there is enough raw material to make the endocannabinoids that the body needs. Excess nutritional building blocks are simply used to make something else. This should be the first step for any intervention of increasing ECS functioning.

DECREASE LIGAND DEGRADATION

If we decrease the breakdown of endocannabinoids, they remain free to bind with receptors. We can prevent the breakdown of the endo-cannabinoids by decreasing their degradation enzymes. There are herbs that are known to prevent the breakdown of both FAAH and MAGL, such as the flavonoids found in red clover, soy, and tea; echi-nacea; and the CBD in cannabis.

INCREASE RECEPTOR NUMBER AND FUNCTION

The more receptors for chemical messengers to bind to, the more responsive the system. There are a number of known interventions that increase receptor numbers, such as maintaining a healthy gut biome, increasing omega-3 fatty acids, exercising, and implementing acute cannabis use.

Aiding ECDS with Intermittent Dosing of Cannabis

Acute use of cannabis kick-starts the ECS. THC increases CB1 density for up to 14 days, increases sensitivity to all ligands of CB1, decreases transport of AEA and 2AG back into cells for recycling, and stimulates AEA biosynthesis. CBD delays the reuptake of 2AG and AEA at the synapse by inhibiting the transporter and FAAH breakdown of AEA.

Chronic use of high-THC cannabis downregulates and desensitizes CB1 and CB2 receptors, especially at the hippocampus. This down-regulation of receptors directly correlates with the number of years using cannabis. Research findings on downregulation are interesting. Downregulation varies by brain region; not all regions are downregu-lated. One hypothesis is that AEA and 2AG may activate different CB1 receptors in the brain and be affected differently by exogenous THC. Epigenetics of the individual also plays a role in downregulation. In some chronic cannabis users, the CB1 receptor regions on the nucleus are hypermethylated (deactivated), but not in other individuals.

Another thing to consider is THC from whole-plant extracts is a partial agonist of CB1 while synthetic THC is a full agonist. Plant preparations have half the downregulation and desensitization as synthetics; you don't develop tolerance as easily from whole-plant extracts as from synthetic THC. 2AG is a full agonist of the CB1 receptor, and THC (natural or synthetic) may interfere with 2AG's binding because when THC is binding to the receptor, 2AG cannot.

One question that comes up when discussing boosting the endocannabinoid system and downregulation is "What about when I'm taking cannabis extracts for other things such as pain? Will I also be downregulating?" The answer is "Yes, and . . ." When we are utilizing cannabis to work with conditions, we know there will be some downregulation, but we also know we will be consuming the plant regularly while we are working with the condition, which essentially replaces the endogenous cannabinoids. Working with specific conditions is different from consuming cannabis to boost the endocannabinoid system. Where we want to be conscientious is when we stop using the cannabis. If we have been using it long term to work with the condition we would then taper down the use over three weeks to allow our own endocannabinoid levels to work back up.

CHAPTER FOUR

PREPARATION AND DOSAGE

Before making medicine, I begin with offering gratitude for the generous gifts cannabis offers and her willingness to share them with us. I also give thanks to all the forces that allow her to grow and thrive: air, sun, water, earth, insects, mycelium, bacteria, and all beings that contribute to her abundance. Finally, I offer thanks to the people who have carried plant wisdom forward and to those who grow and make cannabis medicine available for us today.

Selecting Plant Material

When choosing any plant material to make medicine, including cannabis, use all your senses. Look at the flowers: Do they look vibrant
even when dried? Are they green or even purple — or brown? We
don't want to start with brown flowers. This means the sample is
either old or wasn't dried properly. Next, break open a few flowers.
Do you see seeds? If so, you can take at least 30 percent off the price
(unless the grower has already done so) because seeds account for
30 to 35 percent of the weight of fertilized flowers. Also check for
mold — we don't want mold. How sticky is the flower material? The
stickier the better — stickiness indicates more resin. And finally, the
big test: How does it smell? You should smell the delicious scent
unique to cannabis. If you don't, the batch might not be the highest
quality. The flowers may still be high in CBD or THC, but they do not
have the terpenes you're looking for.

Cultivars

Choosing cannabis cultivars to work with for making medicine is
crucial. Cannabis is an apothecary unto herself. You could spend
your whole life learning different cultivars and which conditions
they work with. Invest the time to get to know different cultivars by
working with them directly; it may take some experimenting before
you are able to formulate the specific medicines you need. Cannabis
in general may work for a particular condition, like anxiety, but not
every cultivar will help. Some cultivars may actually *cause* anxiety.
You can do some reading and research about particular cultivars,
but you can never be sure the Bubba Kush you read about is the
same Bubba Kush you have in front of you.

   If you have the financial resources, you can lab test flowers, tinctures, and oils to learn the percentages of the constituents of the
flowers you are working with. Cannabis from a dispensary should be
labeled with this information — it's one of the reasons you pay more
there. Testing the potency of your tinctures is helpful when you are

trying to produce consistent medicine over multiple batches. Testing will also help you learn how well you are extracting. To be clear, you can make beautiful medicine with reproducible results without testing. The takeaway here is: you want to know the effects of *the specific batch of medicine in front of you* when working with people.

Each of the cannabinoids and terpenes have specific effects within the body. The art and science of formulating is matching the medicine with the condition. Once you know cannabis works with a condition, you can take the next step: finding the particular cultivar. This final step of matching the correct cultivar will go a long way to increasing your effectiveness as a medicine maker.

The Energetics of Medicine

When it comes to healing, people experience a difference between industrially produced medicine and conscientiously cultivated medicine. The environment that cannabis plants are grown in, the nutrients used, the soil, and how the grower interacted with the plant are all important factors to consider when choosing plants for medicine. The environment that a plant is grown in will express itself in measurable ways (such as the terpene profile) and in ways that are less quantifiable but more qualitative: the energetics of medicine. Perhaps you've noticed this distinction when comparing the bottle of elderberry syrup made from a large manufacturer to the one that you or a local herbalist makes. For guidelines on choosing cannabis to work with, see the section on sourcing cannabis for medicine on page 16.

Scientific Studies

The classification of cannabis as a Schedule 1 drug by the U.S. federal government has effectively tied the hands of scientists who've wanted to study the plant's medicinal effects, forcing them to use synthetic analogues of the plant's constituents. For scientists to get approved for funding to research cannabis in the United States, they must apply to the Food and Drug Administration (FDA), Drug Enforcement Administration (DEA), National Institutes of Health

The pharmaceutical industry labels unpleasant or unwanted effects of a medication as "side effects." What does this mean exactly? It really means that the pharmaceutical has unwanted and *uncontrolled* effects. The presence of unwanted effects is an example of our inability to isolate, predict, and control what happens within our multifaceted, interconnected being. Adverse effects become even more dangerous with isolated extracts of any kind, including plant extracts.

(NIH), and the National Institute on Drug Abuse. The application process requires endurance and commitment from scientists, and even if they are finally granted funding, the cannabis flowers they are "allowed" to work with are subpar and well below the standard that herbalists would choose to work with. In addition, the mainstream medical bias toward single constituents and standardization further limits our collective knowledge of the efficacy of whole-plant medicine for specific conditions.

Unfortunately, recommendations for medicinal dosages of cannabis that are in the public record (even from the National Academies of Sciences, Engineering, and Medicine) are sparse. When reviewing information about dosing in scientific studies, it is helpful to keep in mind that whole-plant extracts are for the most part not used, and whole-plant extracts have proven to be up to 330 times more potent than single isolates. As herbalists, we understand the breadth and scope of using the whole plant as a therapeutic tool for healing. Given our understanding of the mechanisms of the endocannabinoid system and cannabis, herbalists can be trailblazers and leaders in the art of healing with this master plant.

Questions to Consider When Buying CBD Products

When you, as a consumer, walk into a store to buy a high-CBD cannabis product, you are entering a wild and mostly unregulated arena. You can't be sure that what the manufacturer says is in the product is actually in there or at the quantities they report. There are currently no regulations or accountability for the labels "full-spectrum" or "whole-plant extract" on CBD products. This will likely all change in the future, but for now the question from many consumers is, "Where do I go to get quality medicine?" My first answer is to make your own. Some people don't have the time or desire, and that's okay. If you don't make your own, buy from a local herbalist who makes tinctures or infused oils and who knows many plant medicines (preferably one who isn't trying to sell you one product to "cure" everything). If you don't have access to an herbalist, go to a local grower who extracts or knows how the flowers they grew were extracted so you can ask questions. If you choose to buy from a bigger company, you can ask these same questions.

You want in your medicine everything the plant made while she was growing: the cannabinoids, terpenes, flavonoids, fats, waxes, and chlorophyll (some extractors remove one or more of these). You want these things from the plant itself, not added in from another source (terpenes from other sources, for example, are sometimes added). If it doesn't *smell* like cannabis, it doesn't have terpenes (yes, your gummies should smell like weed). Finally, if you want to be part of changing the dominant paradigm, buy from growers directly, pay a fair price, and buy from women and people of color.

When you buy from anyone, ask these questions to make sure you are getting the best quality:

Are you using organic flowers to make medicine?
You want organic flowers.

How are you extracting?

If the answer is "with butane," move on. Ethanol or supercritical carbon dioxide are the industry standards. Home medicine makers usually use ethanol or butane, but I advise against using butane.

Do you test your end product or flowers for heavy metals? Pesticides? Mold?

You want medicine free from heavy metals, pesticides, and mold.

When you extract, what are you taking out and what are you leaving behind?

Sometimes the flowers come in a little "hot," containing more than the legal 0.3 percent THC. Some companies remove enough THC to bring the level below the legal limit, but you want to have 0.3 percent because you need some THC. You actually want equal parts THC and CBD for the best medicine, but that's not legal everywhere. If they take anything else out, you don't want the product.

Does it smell like cannabis?

You should be able to smell the terpenes. If it doesn't smell like cannabis, it's made with isolate and you don't want it.

After you've extracted, are you putting anything else back in?

If they have, you don't want the product. Companies often use additives to compensate for poor-quality starting flower, most commonly CBD isolate. That way they can use poor-quality flower and call it full-spectrum or whole-plant medicine while adding in some isolate to make it more potent (and charge you more money!).

Do you have "isolate" in your product?

If the answer is yes, you don't want it.

Methods of Intake

There are two primary ways people ingest cannabis: inhalation and oral ingestion. A person can inhale smoke or vapor; tinctures or oils (and things made with the oils) can be ingested orally. The oral method is preferred for all conditions, while inhalation is best for breakthrough symptoms.

Inhalation

The quickest and shortest path into the bloodstream is through the lungs. When smoke or vapor is inhaled into the alveoli of the lungs it can pass easily into the blood.

Methods for inhalation include combustion, which is igniting the cannabis flower and breathing the smoke through a pipe, joint, or water bong; vaporizing the flower with a vaporizer; or inhaling the vapor of a resin extract ("dab," "shatter," and "wax" are a few names). For medicinal purposes in general, with only two exceptions, inhaling resin extracts is not recommended. Don't let anyone tell you different. This potent medicine causes an extreme increase in cannabinoid levels in the blood followed by a quick decline, requiring frequent multiple dosing. All conditions except extreme pain are better served by a slower and prolonged increase. Combustion is low tech and inexpensive; rolling papers cost only a few dollars. You can also spend a few hundred dollars on a handcrafted, blown-glass water pipe or bong. One downside of combustion is the pulmonary irritation that comes with inhaling toxic metabolites. Water bongs can remove some toxins (including nitrosamines, ammonia, acetaldehyde, benzene, and carbon monoxide), but they do not remove polycyclic hydrocarbons.

Vaporization occurs at a lower temperature than combustion and eliminates all toxins, including hydrocarbons.

BENEFITS AND POTENCY

Inhalation in conjunction with oral ingestion is suggested for breakthrough symptoms experienced in pain and anxiety. For example, if you have pain with osteoarthritis, you might overwork the painful knee joint on a particular day, resulting in more pain. Inhalation is not recommended for dosing throughout the day. The length of bioavailable time is much shorter than with oral administration, thus it is not recommended as a standard method of dosing; individuals would need to dose often to maintain blood levels of the necessary cannabinoids. All cannabinoids and terpenes are available in vapor or smoke, but no flavonoids are available to the body with inhalation.

Inhalation provides a medium-strong dose that can be easily administered with an onset of effects within 5 to 10 minutes. This short onset time allows dosages to be easily regulated. Peak blood concentration occurs within 30 minutes, and after 3 hours the blood level is below 5 nanograms per milliliter, the legal limit in the state of Colorado. (Nanograms per milliliter, abbreviated ng/mL, is a unit of measure commonly used to express drug levels in urine and saliva. A nanogram is one billionth of a gram.)

Studies have shown the efficiency of THC delivery through different inhalation methods. Hand-rolled cannabis cigarettes, or joints, yield 27 percent of the available THC; pipes yield 50 percent of the available THC; and water bongs yield 10 to 20 percent. Interestingly, though the delivery methods differ in efficiency, resulting THC blood levels are all about the same. Individuals unconsciously self-regulate the THC level when inhaling, regardless of the method of inhalation.

Oral Ingestion

Cannabis for oral ingestion comes in tinctures, oils, capsules, edibles, infusions, and teas. Oral administration is the best method of ingestion for treatment of all chronic conditions. Oral administration lasts 5 to 8 hours, with a peak concentration from 1 to 8 hours (average 4 to 6 hours). Dosing every 6 hours keeps cannabinoid levels constant, rather than fluctuating as they do when inhaling. The half-life of ingested cannabinoids is 18 to 32 hours. This means, for example, that 24 hours after a person ingests 10 mg of THC in an edible, she will still have approximately 5 mg of THC circulating in her blood. After a dose of 5 to 10 mg, it takes about 4 hours for the THC level in the blood to drop below 5 ng/mL, the legal limit in the state of Colorado. Generally, 6 to 20 percent of ingested cannabinoids become bioavailable.

The onset of effects after oral ingestion can take anywhere from 15 minutes (tinctures) to 3 hours (edibles), depending on stomach contents and liver function. This delay can be problematic for people who ingest cannabis orally and get impatient waiting for the effects (many people have a story about cannabis edibles). A good rule of thumb is to wait at least 3 hours to see if effects occur. If, after 3 hours, effects are not felt, administer one-fourth as much as the original dose and wait another 3 hours.

COMMERCIAL PRODUCTS AND BIOAVAILABILITY

Companies trying to get you to buy their cannabis products will often use "sciencey" words with questionable physiology. Product "bioavailability" is one of these. If you make a tincture with even 6 percent bioavailability, you have good medicine, especially because you are making it yourself, know the origins of the flowers, and are likely starting with a better plant source than big companies do. The fact that you are making whole-plant medicine ups your potency anywhere from 4 to 330 times over the commercial products. Sublingual drops or buccal sprays are simply a marketing device and not necessary for medicinal purposes. They're absorbed slightly faster and avoid first-pass liver metabolism, but neither reason makes their product better than what you would make. Period.

BENEFITS AND POTENCY

In studies comparing tinctures and capsules (including edibles), tinctures surpassed capsules in bioavailability, consistency of absorption, and consistency of THC delivery. Unfortunately, these studies were conducted with isolated synthetic THC.

Tinctures and oil infusions incorporated into edibles are all absorbed by the digestive system. Fat will need to be added to infusions and teas because cannabinoids are fat soluble. The terpenes and flavonoids in the cannabis are retained if the extraction process remains below 100°F (38°C) degrees.

A standard place to start for oral dosage is 2.5 mg THC. Work up to 10 to 15 mg THC if needed. Between 10 and 15 milligrams of THC is considered a "sweet spot" for most conditions, especially pain. At 10 to 15 mg, you are receiving the maximum benefit per dose. Upping your dosage does not deliver more relief; it will just cost you more money without much benefit.

Medicine-Making Principles

When making medicine with cannabis, it's important to keep a few points in mind:

* Cannabinoids are lipid soluble. So, if you are extracting with water extractions (for infusions or tea), you need to add fat (like coconut milk or half-and-half).

* Cannabinoids need to be decarboxylated to be activated. If you want the acid form, you would not decarboxylate. You are making an informed choice about what constituents you want in your medicine. Once you know what constituents you want, you can decide whether to decarboxylate or not.

* Flavonoids are available only in unheated preparations.

* Terpenes are lost when heated above 70 to 100°F (21 to 38°C). Decarboxylation with heat will destroy many of the terpenes. If you can smell the cannabis as you are decarboxylating, you are losing terpenes; if you can smell cannabis in your medicine, you have kept some of them.

* If you overheat cannabinoids or heat cannabinoids for too long, they will convert to CBN and cannot be reactivated; there's no coming back from CBN.

Maximize Plant Material Surface Area

When making a tincture (or infused oil), the more surface area of the plant you can expose to heat during decarboxylation (more on this process below) and to alcohol or oil, the more constituents will be extracted and the stronger the medicine. The cannabinoids and terpenes are in the trichomes. The more trichomes exposed to the tincture or oil, the stronger the medicine. So you will want to chop up the cannabis plant. A few simple methods:

- Grind dried plant material in a coffee grinder; this is good for small amounts. Dedicate the coffee grinder to cannabis; otherwise your coffee will smell and taste like cannabis.

- Grind dried plant material in the dry attachment of a Vitamix blender; this is good for larger amounts.

- Run the flowers through a #10 sieve.

- Send large quantities of plant material through a wood chipper (*Fargo* style!); this would be for very large quantities.

I've found that using a Vitamix is the easiest and fastest method. A tip on cleaning the Vitamix or sieve: rinse it with organic grain alcohol when you're done, then use that same alcohol for making your tincture. You get two benefits in one: you clean your equipment and recover any resin left behind on the blades or screen.

Standardization

If you are making medicine and want consistent dosing across multiple batches of medicine, a few additional steps can help.

- Grind flowers to a similar consistency each batch; I use a dry attachment on my Vitamix, or you can use a #10 sieve.

- Weigh the flowers in grams.

- Measure the alcohol in milliliters.

- Decide on the ratio of flowers to alcohol, and use that ratio consistently. Remember: grams of flowers to milliliters of alcohol. A ratio of anywhere from 1 gram of flowers to 5 milliliters (mL) of alcohol, down to a 1:10 ratio, is a good range. A 1:5 ratio would yield a more potent extract. For example, 114 grams of flowers to 570 mL of alcohol (2.4 cups) would be a 1:5 ratio. And 114 grams of flowers to 1,140 mL of alcohol (4.75 cups) would be a 1:10 ratio.

◆ Knowing the weight of flowers and milliliters of solvent will allow you to calculate dosage (more about this at the end of this chapter).

THE MYTH OF FREEZING

You may hear people suggest an additional step of putting the alcohol and the ground plant material in the freezer for 24 hours separately, combining them, and then putting the mixture back in the freezer to macerate. The reasoning is that less chlorophyll will be extracted. This is true — but it is unnecessary and undesirable for tincture making. This is a matter of aesthetics in making resin extract to smoke (dab, shatter, or wax). People who make dabs are looking for a golden color, and they think chlorophyll gives an unpleasant taste. Another reason is that freezing reduces unwanted lipids and waxes in finished extract. So what? We want full-plant herbal medicine! Everything the cannabis plant has to offer. Have you tasted valerian? Hops? Kava? We're not making a delicious drinking beverage; we are making medicine.

What Is Decarboxylation, and Do I Need to Do It?

Within the trichome there's an entire world of chemical reactions occurring. Metabolites are being made for the cannabis plant's survival. The cannabinoids we are hoping to use are found in their acid form (THCA, CBDA, and CBCA, discussed in chapter 2) within the trichome of properly harvested plants. While the acid forms have medicinal properties, the "active" forms of the cannabinoids — THC, CBD, CBG, CBC, THCV, CBDV — have the acid removed. The process of removing the acid is decarboxylation. The acid removal occurs in two ways: over time or with the addition of heat. If you were to make a tincture or oil preparation and let it sit for a year, decarboxylation would begin to occur. But most people do not have that kind of time, so they use heat.

Cannabinoids in their active form will further oxidize into the inactive form cannabinol (CBN), also through heating or the passage of time. Real medicine-making skill is in decarboxylating the acid form to active cannabinoids without taking them all the way to CBN. Once converted to CBN, you cannot go back. Anyone who has had cannabis for years knows that at some point, just by sitting in the jar, it loses potency. You may or may not get the euphoric feeling with aged cannabis, but you will definitely get the sedation. Freezing cannabis does not slow this process down; time still exists in the freezer.

When you vaporize or combust a dried flower, you are decarboxylating and creating the active form of the cannabinoids right on the spot — and getting terpenes in the vapor or smoke as well. If you were to make tea with hot water poured over your flowers, you are also adding heat and decarboxylating — but you will need to add fat to the infusion for better absorption into the body.

Heat Decarboxylation

Home medicine makers can use either an oven or a turkey roaster for decarboxylation. The easiest method for high-CBD or high-THC strains, or mixtures of both, is to put the sieved or Vitamixed flowers in a half-inch layer on a cookie sheet in the oven or the bottom of the turkey roaster and heat at 250°F (121°C) for 80 minutes, stirring every 15 to 20 minutes. THC and CBD decarboxylate at slightly different temperatures, but this method will safely extract both cannabinoids without converting to CBN. As you get to know your oven and your own grinding process, you can fine-tune your method by heating a bit longer or at a higher temperature.

Don't trust your oven thermometer. Buy a separate thermometer to put in the oven to make sure you are at 250°F (121°C). Heating above 100°F (38°C) evaporates terpenes; you will smell them evaporating as you heat the flowers. Adding a cover to the pan in the oven will help retain some of the terpenes. Be sure to scrape all the "dust" (keif) from the bottom of the pan into your jar. That dust is the trichomes that have broken off the leaves and flowers. You want that!

Mini Decarboxylators

You can buy a handy-dandy mini-decarboxylator that will decarboxylate an ounce of flowers in less than 2 hours, leaving the terpenes intact. The only drawback is that the machines decarboxylate only 1 ounce of flowers per batch.

Terpenes and Heat

Terpenes are an essential part of your medicine. If you want to retain the terpenes, do not use heat above 100°F (38°C), or do use a mini-decarboxylator that retains terpenes. If you want the acid forms of the cannabinoids and the terpenes, make a tincture using dried or fresh material and use it within one year. After that, the acids begin to convert to the active constituents regardless of whether they've been heated or not. It may take another full year to completely decarboxylate to the active forms. That is, it can take two years for fresh flowers to decarboxylate.

Fine-Tuning

As you make medicine over time, you will develop your own method for grinding the flowers and somewhat standardizing the whole process. If you want to see how well you are extracting and decarboxylating, make your tincture and have it tested. The ratio of acid to activated constituent will tell you how well your decarboxylation/extraction process works. The more THCA or CBDA you have left over, the longer you need to decarboxylate. After a few rounds of making medicine and testing, you can fine-tune the process. Maybe you'll need to decarboxylate for a little longer in the oven, or maybe go to a slightly higher temperature.

We don't have a lot of data regarding the amount of time for decarboxylation at room temperature (71°F [21°C]). Data from a 1978 study on the rate of decarboxylation over five years for dried flowers shows that, at around a year, decarboxylation begins and reaches maximum THC levels. The levels plateaued for another year, and then began to degrade further into CBN after those initial two years. We can assume CBD will behave in a similar fashion.

The same study looked at the constituents in alcohol (which are probably similar in oil): THC begins to degrade somewhat at three years. It's interesting to note that in a 1972 study, a 43-year-old extract tested positive for CBD, THC, and of course CBN.

BENEFITS OF COMMERCIAL EXTRACTIONS

One major benefit of industrial-scale supercritical carbon dioxide (CO_2) or ethanol extractions is potential retention of the terpenes. I say "potential" because you need to ask if the terpenes extracted from the plant material were put back into your tincture or oil. Ask exactly that question. Why? Not all ethanol extractors have a terpene catch, so they lose terpenes just like you would in your oven. Second, many manufacturers sell the terpenes they extract separately and sell the oil or tincture without the terpenes to the customer. Given current labeling practices, a tincture or oil without the terpenes can still be called "full spectrum" or "whole plant."

Acids, Actives, and Terpenes

To be clear, you can make beautiful medicine by decarboxylating to obtain the active, nonacid forms THC and CBD (losing a large portion of terpenes) and then making tincture or oil. You can make beautiful medicine by not decarboxylating and retaining the acid form and all of the terpenes. If you want the best of both worlds and you don't own a fancy mini-decarboxylator, you can make a tincture of the decarboxylated forms and another tincture with the acid forms and terpenes, then combine the two! The possibilities are endless. For most conditions we address in this book, the dosage recommendation is for decarboxylated THC and CBD.

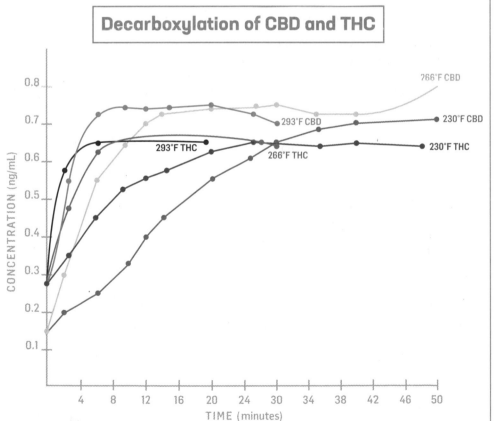

Forms of Medicine

The three main forms of medicine we will make are tinctures, infused oils, and resin extracts. Most home medicine makers can get the medicines they need from tinctures and oils.

General Guidelines for Tinctures and Infused Oils

* If you are using trim, the ratio can be 1:5 — that is, 1 gram of dried trim per 5 mL of oil.

* If you are using flowers, you can use a higher ratio, up to 1:10.

* Flowers tend to be more potent than trim because they have more trichomes — but trim works just fine for medicine making.

* The traditional method employed by herbalists is to fill a container with plant material (size matters), then fill the container with oil or alcohol; this usually ends up at around a 1:5 ratio.

Tinctures

Tinctures of cannabis are suitably potent for almost all conditions except for some seizures and fighting cancer. They absorb quickly into the body (in some cases as quickly as 10 to 30 minutes) and can be easily made by the home medicine maker.

Based on the very few studies out there, it takes two years before tinctures start converting to cannabinol (CBN). Most herbalists like to make medicine to last a year — and then you get to make more!

What It Involves

Making a tincture is the process of extracting active constituents from a plant with alcohol. Tinctures are one of the main medicines herbalists use. The process can be as simple as cutting plant material into small pieces, filling a jar with the plant material, and covering the plant material completely with alcohol. Then, as we herbalists say, "Shake and pray every day" for three to six weeks, strain, and store in a cool, dry place. Yes, it can be that simple.

Actually, it doesn't take three to six weeks for cannabinoid extraction. It only takes an hour! (I found an unpublished study that showed extraction after 30 minutes.) Yes, I just said an hour. It's really true, and I know that's hard to believe. We herbalists are set in our ways, and if you want the energy of a full lunar cycle, wait three to six weeks. It will not make a less-potent medicine (most of us would say it's *more* potent with that full lunar cycle).

Any process that breaks the cannabis trichomes and exposes their contents to alcohol will work for making a tincture.

TINCTURING WITH ALCOHOL

When making a tincture with something as resinous as cannabis (like calendula or spruce tips or poplar buds), an alcohol content of 95 percent or higher is recommended. Studies done with lower alcohol percentages show diminished extraction rates for active cannabinoids. Organic grain alcohol bought in bulk is the cheapest route. Organic, in this case, means it's made with a carbon-based mash such as corn or another grain, potatoes, or sugar. It does not mean it meets USDA organic growing standards. There are, additionally, grain alcohols that are also labeled organic, and are organically grown, such as organic grape alcohol. Prices for these are two to three times higher than regular grain alcohol. Note that organic grain alcohol is a distilled alcohol and is naturally gluten-free regardless of the mash or marketing.

TINCTURING WITH GLYCERIN

Some people prefer not to use alcohol for sensitivity or substance-abuse reasons. You can extract using glycerin instead. Vegetable glycerin is derived from fatty acid esters in coconut, soy, or palm oils. Vegetable glycerin extracts about one-third as well as alcohol, so if you use the same amount of plant material in vegetable glycerin as in alcohol, the medicine made with glycerin will be one-third as potent as the alcohol extraction.

Step-by-Step Tincture Making

1. Grind flowers.
2. Decarboxylate if you want nonacid forms of cannabinoids.
3. Add alcohol.
4. Shake and pray at least 1 hour (you can also just let it sit, shaking every 15 minutes).
5. Strain out the plant material.
6. Squeeze or herb-press the plant material to extract any remaining tincture, then discard or compost the plant material.
7. Let sit overnight.
8. Strain the next day through a coffee filter to remove all fine particulate matter left in the tincture.
9. Label and store tincture in a cool, dark space.

CANNABIS-INFUSED HONEY

A true gift from the bees, sweet golden honey is an important addition to our medicine chest. Unfortunately, honey does not extract cannabinoids due to its hydrophilic (water-loving) nature. Cannabinoids, remember, are lipophilic (fat loving). If you infuse cannabis flowers in honey it will smell like cannabis because you are extracting some of the terpenes but not enough of the cannabinoids. But don't disregard honey! One option is to powder cannabis flower, add it to the honey, and blend it into a paste.

Botanical Extractor Machines

If you're going to be making lots of tinctures or oils, invest in a botanical extractor machine. Buy direct from the company to obtain the warranty (if you buy from an online retailer, you might lose the warranty option). Add decarboxylated plant material to the maker, add alcohol, set the temperature to the "no temperature" setting for 1 to 8 hours (1 hour will work just as well as 8), dance to the disco lights of the machine, come back in the allotted number of hours, and proceed with the final steps of making a tincture. Important: Don't skip the separate decarb-in-the-oven step and just turn the heat on in the extractor. I did that and the alcohol evaporated, risking a fire. Luckily, I checked on it before it completely destroyed my machine or caught on fire.

Infused Oils

When plant material is placed in oil and heated, constituents will extract just like in a tincture extraction. Saturated fats like coconut oil, fractionated coconut oil (MCT oil), butter, and ghee are excellent media for cannabis. These saturated fats extract cannabinoids better than unsaturated fats like olive and sesame oils. Depending on which oil you choose, the cannabis medicine you make can be taken internally (the oil you use for these must be edible) or used topically. Cannabis-infused coconut oil makes great edibles as well as a good massage oil for topical use.

CBD Oils on the Market

CBD oils on the commercial market are made by extracting with either carbon dioxide or ethanol, evaporating out the solvent to get a resin, then diluting with olive oil or fractionated coconut oil, commonly called MCT oil (for medium-chain triglyceride). They are not the infused oils we herbalists know and love. Why go through the trouble of extracting a resin only to "dilute" it with MCT or olive oil when you could just make an infused oil? The answer is plant quality. Commercial large-scale growers and extractors can use lower-quality flowers (8 to 10 percent CBD on average) because they put the whole plant, including stalk, leaves, and flowers (termed "biomass" in the industry), into a grinder and extract all the way down to a resin, concentrating the CBD. Then they dilute it to whatever percentage necessary for the final product.

Herbalists make medicine from the plant parts known for medicinal qualities. We use the flower and the leaves containing trichomes only. We don't have the luxury (or desire) to use large quantities of low-quality plant material to extract from. Our model is to use the highest-quality plant material harvested at the peak time for making medicine.

Making an Infused Oil

When we heat oil, we don't have to worry about it evaporating or igniting, so decarboxylation as a separate step *could* be eliminated. In the traditional herbalist sun-infused method: start with chopped or ground dried plant material. Place the plant material in a jar, add enough oil to fully cover, place the jar in a sunny spot, then shake and pray for 3 to 6 weeks. How well extraction occurs depends on the temperature the oil reaches during the day and the number of days of extraction; this particular method of oil infusion has the most variability of potency and extraction. Based on my experience, decarboxylating the flowers first yields the most potent medicine. If you would rather not, it is important to extract for longer with the plant in the oil.

If you aren't going to decarboxylate first, try any of the following variations on the traditional method to make potent and more consistent medicine. All of these processes will work. The variable factor is figuring out how long to heat the flower in the oil. It's difficult to know if you are getting the internal temperature of the flowers up to 250°F (121°C). It's likely that you are not, so longer heating times are necessary. The most consistent and potent process I've found is to decarboxylate the plant material in the oven first, then put it in a botanical extractor, cover with oil, and set to 130°F (54°C) for eight hours.

◆ Place plant material in a botanical extractor and cover with oil. Set temperature to 130°F (54°C), the lowest temperature, for 8 hours. The benefit of using this machine is that it heats and stirs automatically and you retain terpenes. The time and temperature will vary depending on the kind of oil you use. When using olive oil, for example, it is better to keep the temperature lower than when using coconut oil, a higher-heat oil.

- Place dried plant material in an oven-safe pot. Cover plant material with oil. Place the pot (with a lid) in a preheated oven at 270°F (130°C) for at least 4 hours. Remember, you want the oil to reach 250°F (121°C) for 1 hour.

- Place plant material in a double boiler on the stove top, cover with oil, and then cover the double boiler. Heat and stir frequently for at least 4 hours. Again, heat the oil to 250°F (121°C) for 1 hour.

- Place plant material and oil in a slow cooker for at least 6 hours set on 250°F (121°C). Stir every hour.

Usually folks worry about "overcooking" their plant material and converting to CBN. I've mistakenly decarbed plant material in the oven for 8 hours and had zero conversion to CBN.

Whatever method you chose, at the end of the allotted time, strain the plant material from the oil, then press it out with an herb press or squeeze in cheesecloth. Compost the plant material and retain the oil. Place in a cool dry place, in the refrigerator, or in the freezer, depending on the storage requirements for the type of oil you used.

HERB PRESSING

For years, I simply used my hands to squeeze plant material and extract all the goodness I could from the plants. This works just fine, especially if you're not investing lots of money into the medicine you are working with. But because working with cannabis can be costly, it might serve you well to press the tincture or oil more efficiently. Recent studies found that pressing the material increases the CBD extracted by 1 to 2 percentage points. A stainless-steel fruit wine press can do this.

Resin Extracts

The most potent medicine herbalists can make are resin extracts. At dispensaries, resin extracts are a complete line unto themselves and are becoming a favorite in the personal-use world. These concentrates are also commonly called dabs, shatter, wax, or full-extract cannabis oil (labeled FECO, this can potentially be confusing because it is not an infused oil). Resin extracts are obtained by extracting cannabis constituents in a solvent and then evaporating the solvent. The final product is a sticky, resinous extract that contains 50 to 90 percent cannabinoids.

BENEFITS AND POTENCY

Concentrated resin extracts are recommended for extreme conditions such as postsurgical pain, breakthrough seizures, or fighting cancer. Potency can be very high: as much as 90 percent THC or CBD. With potency levels this high and the fact that resin extracts are so quickly bioavailable when inhaled, caution should be employed. Tolerance can develop quickly, and the danger of addiction rises. Additionally, resin extracts require a high level of chemical processing and, depending on the solvent used, might contain adulterants and toxins. Finally, because resin extracts are highly concentrated, they will also concentrate any toxins, including heavy metals from the ground the plants were grown in. Cannabis is a known bioaccumulator.

The recent introduction of vaporizers for resin extracts has completely changed the inhalation method for cannabis use. Before these specialized vaporizers were created, there wasn't a great way for people to smoke the gooey resin. Now people can take one "hit" (inhalation) of the resin extract and receive a much higher dose of THC than when taking a hit of the flower. A hit of a 15 percent THC flower will deliver approximately 2.5 mg of THC. A hit of resin extract could deliver 5 times that amount.

Concentrated resin extracts can also be taken orally. Absorption through the gut lining is dependent on the same variables as other oral methods (how full the stomach is and what kind of food is in the GI tract). A resin extract can be hard to swallow, literally. A common practice is to put the dose in a gel cap for ingestion.

Making a Resin Extract

Before you begin this process, please read through all the steps and get all your materials and equipment prepared. You can expect to get approximately 100 to 200 mL of resin from a gallon of tincture, depending on whether you use trim or flowers, how resinous they are, and how far down you evaporate.

1. Weigh the plant material to be used (for dosage calculation).
2. Decarboxylate.
3. Make a tincture in your botanical extractor or the traditional way.
4. Strain with large strainer.
5. Strain through paper coffee filter (an important step!).
6. Do this next step outside! Place strained tincture in a rice cooker with the lid off or in a waterbath in a crock pot. Alcohol is extremely volatile. Do not breathe the vapor. Aim a fan over the cooker to allow good air flow. Allow the alcohol and water to evaporate off (this usually takes an hour or less). When the liquid is reduced and starts to thicken, *do not walk away*. Periodically remove the container from the heat source and swirl the ingredients, being careful to not inhale the vapor. Bubbles will get smaller as the process comes to a completion. Remove from heat source and keep swirling.

   CAUTION: Alcohol is extremely volatile and boils at a lower temperature than water: 173°F (78°C). This means the alcohol will boil off first. I recommend adding a few drops of water to your mixture as the alcohol is boiling off to protect the resin from scorching during the final evaporation. Be sure you do this procedure outside where there is plenty of ventilation. Good ventilation and lack of an open flame (that's why we use an electric burner or rice cooker) prevent the alcohol vapor from igniting. Even if you are outside, and the wind isn't moving the vapor away from your pot, you still risk ignition. Good ventilation also protects you from inhaling the fumes.

7. If needed, use a coffee-mug warmer to finish. Keep the mixture warm and draw the liquid up into a syringe. (Large, non-needle syringes can be purchased from a pharmacy or online. They are the ones used to give children oral doses of pharmaceuticals.)

8. Optional: Don't evaporate all the way to full sticky resin; instead, when you've evaporated most of the alcohol off, pour it into a measuring cup, noting the amount, and then dilute it with MCT or olive oil. A gallon of tincture will evaporate to roughly 120 mL of resin. Pour the resin into a Pyrex measuring cup and add an equal part MCT or olive oil. I prefer doing this because it makes the resin easier to work with, it's less sticky because it's mixed with an oil, and it prevents the resin from burning (yes, I've done that). Because this medicine is used orally, I'm not concerned with removing chlorophyll or trying for a golden color. And because I know how much plant material I started with and what my final volume is, I can calculate dosage.

9. Measure the final volume for dosage calculation.

This may all sound straightforward, simple, and clean. In theory, it is. In practice, not so much — resin is very sticky and can get all over you and your countertop and containers. You will develop methods for dealing with this problem. Warm resin is easier to work with, for one, as it flows much more easily than cold resin. And do not lick your fingers! This is very potent medicine; a drop can contain 14 mg of THC!

Note: Undiluted grain alcohol is a great solvent for cleaning up.

Resin-Making Machines

Several commercially available products can create a resin extract in an enclosed container, under pressure, using alcohol as the solvent. These machines lower the risk of ignition, are easy to clean up, and recoup up to 75 percent of the alcohol for reuse. The downsides are the cost (machines start at $500) and that they can only make small quantities of resin at one time.

Extracting with Butane

Butane is efficient, inexpensive, and easy to obtain, which makes it an attractive candidate for extractions. But extracting with butane is dangerous and not advised; butane is a carcinogen and cannot be completely purged from the final product. Studies have found butane levels in so-called "purged" (or clean) extracts that are equal to the level of terpenes.

Extracting with Supercritical Carbon Dioxide (CO_2)

Many big operations use supercritical CO_2 for extraction. The first extraction is done below freezing temperatures, which preserves the terpenes. This method is, however, unaffordable for the home medicine maker.

Dosage

Beginning herbalists work with doses of a cup or a quart for an infusion or decoction (infusion by boiling rather than steeping) or one to two droppers of tincture. In most cases, plants and herbs are very safe, and experimenting with dosage will pose no risk of major ill effects.

Some plants, however, are toxic or have intense effects. We designate these as low-dose botanicals. Low-dose botanicals are taught to herbalists later in their course of study, as they more fully understand then how plant medicine works within their own bodies. Low-dose botanicals are administered in small doses because of their potency; you typically need only one to three drops of medicine.

Low and Slow

While cannabis is not toxic, I consider it a low-dose botanical because even a little bit can produce an altered mental or physical state. One cannot die from ingesting too much cannabis or too much THC, but cannabis can impact one's daily functions at work or at home. Starting with low doses is advisable as the body adjusts. Too large of a dose could deter someone from taking the medicine in the future (not everyone likes to feel altered).

We can ease people into medicinal cannabis by calculating the potency of the medicine we make. Cannabis from a dispensary should be labeled with potency information. If you're growing it yourself, I recommend having a lab test the flower's potency.

People develop tolerance to botanicals; to achieve the same effects over time, higher doses need to be consumed. When taking the minimum effective dose, tolerance can develop in two weeks.

Minimum Effective Dose

What is the smallest dose that will produce a desired result? Start with one drop of tincture and wait one to two hours for effects to set in. If you don't feel the anticipated effects, take another dose. Wait. Once the optimum oral dose is found, take it, on average, every 6 hours to maintain blood levels.

I know the idea of "minimum effective dose" can be frustrating; some people want a specific measurement. However, people and plants are organic beings; no two are exactly alike. As practitioners, we do our best to get to know both the person and the plant, then make educated decisions about the minimum effective dosage for each individual. Working with plant medicine is both an art and a science. The art of herbalism requires us to learn about and honor the nuances of each person and each plant, and to strike a balance in each formula we create. The science of herbalism is cultivated by educating ourselves about the plants, proper methods for extraction, and application of our knowledge to each unique situation.

For all conditions, oral dosing is preferred, at a minimum effective dose every 6 hours to alleviate symptoms. Oral dosing provides sustained levels of cannabinoids in the blood. Inhalation methods can be used for breakthrough symptoms as they occur, but bloodstream cannabinoid levels spike and fall off quickly with inhalation.

All doses are calculated in milligrams per milliliter (mg/mL). If you want standardized, repeatable medicine, keep careful notes on your methods and measurements.

Cannabis-Naïve vs. Cannabis-Savvy

People new to cannabis as medicine will likely have a low tolerance and may feel effects at a very low dose of 1 milligram or lower. Always start with a low dosage when starting a new round of medicine. Even people experienced with cannabis should start at a low dose with new formulas or a new version of a formula they've been working with.

Balance the THC-to-CBD Ratio

For all conditions except seizures, a balanced 1:1 ratio of THC to CBD is a good place to start. CBD modulates the negative effects of THC, including tachycardia, anxiety, mental impairment, and memory impairment. Working together, CBD and THC have anticancer actions, help decrease inflammatory markers and pain, and potentiate chemotherapy and opiates more than either constituent alone. Balanced cannabinoids also lower the desire to use more. For people new to cannabis, it is advisable to start with strains higher in CBD before using strains higher in THC.

How do we measure a 1:1 ratio? The only way to know this information is to have your flowers or medicine tested in a lab. Even if you are told that Ghost Train Haze, for example, has a potency of 28 percent THC, you can't know the potency of the flower *in front of you* for sure until you have it tested. Test results will give you the percentage of each of the cannabinoids within the flower. In an ideal world, the flowers you use for medicine would have equal parts CBD and THC.

But it's not an ideal world. Not to worry — we herbalists know how to combine different tinctures to make medicine. We can combine our high-THC Bubba Kush with high-CBD Cherry Wine to make a 1:1 tincture. It just requires some math and careful measurements. Knowing how to do this enables us to be infinitely creative with our recipes. When we work with and understand one particular cultivar high in THC, we can pair it with high-CBD cultivars to formulate for different conditions. That's where the fun is!

I like to have my flowers tested, make my medicine, calculate potency, and then have the tincture tested once in a while to check and see if I am extracting as well as I could be. Based on testing my own tinctures, I add a fudge factor of 20 percent when calculating potency. It's tough to decarboxylate *and* extract all the acid forms of constituents. We try for 100 percent, but it's usually more around 80 percent of what the math says.

How to Calculate Potency and Dosage

We can calculate the potency of our tincture, infused oil, or resin extract with some straightforward math. You must know three things: the potency of your flowers, the weight of the flowers in grams before you start making your medicine, and the volume in milliliters of infused oil or tincture at the end.

1. Measure in grams the plant material you will use.
2. Convert grams to milligrams by multiplying by 1,000.
3. Calculate how much THC or CBD you have in all of the plant material. For this, you need a lab test to tell you the percentage of cannabinoids. Multiply the milligrams of plant material by the percentage of cannabinoids from the lab (THC or CBD). Calculate each individually.
4. Make your tincture, infused oil, or resin extract. Strain.
5. Measure liquid in milliliters.
6. Divide milligrams of constituents by milliliters of liquid.

A note about decarboxylation and extraction: Full decarboxylation and extraction rarely occurs; some of the acid forms remains in the tincture, and some of the decarboxylated material remains in the plant material. Until you test for actual levels, we will use the fudge factor of 20 percent in our calculations. The ultimate goal is full extraction and full decarboxylation, but until then we use the fudge factor.

THE SIMPLARS METHOD

I want to pause here and mention that while testing and calculating your medicine's potency are good and informative and help create reproducible products, you *can* make beautiful medicine that helps people without testing or calculating. Our herbal ancestors made amazing medicine without fancy lab equipment. You can develop a standardized process that delivers approximately the same potency in each batch of medicine you make.

Math for Medicine

If you want reproducible medicine, it's important to calculate dosage. To do this, you'll need a scale and a measuring cup with metric measurements. There are a few laws of measure, like gravity, that can't be changed, and they need to be remembered for calculations. Finally, since we don't always completely decarboxylate nor do we completely extract everything out, we will use a fudge factor of 80 percent (we probably only get 80 percent of what is possible).

VOLUME

- 30 milliliters (mL) = 1 ounce
- 1 mL = 2 droppers of a 1-ounce bottle
- 1 dropper = 18 drops of a 1-ounce bottle

WEIGHT

- 1 ounce = 28 grams
- 1 gram = 1000 milligrams (mg)

The Math

1. Give thanks.
2. Weigh plant material in grams (know percentage of CBD or THC).   ___.__ g
3. Convert to milligrams (multiply by 1,000).   _____ g x 1,000 = _____ mg
4. Multiply % constituents by milligrams of dry plant material (this tells you how many mg of CBD or THC you have in all the plant material you are using).   _____ % x _____ mg
5. Make medicine, strain, and measure liquid in milliliters.   _____ mL
6. Divide milligrams of constituents by milliliters of liquid.   _____ mg/_____ mL
7. Multiply by fudge factor of 80%.   _____ mg/mL x 0.8% =

_____ mg/mL

They're laws; you can't change them. Just memorize them.

VOLUME

1 fluid ounce = 30 milliliters

1 milliliter (mL) = 2 droppers full of a 1-ounce dropper top (A "full" dropper isn't necessarily full all the way up to the bulb. It's where the fluid naturally stops; usually about halfway up the dropper.)

1 dropper = 18 drops

WEIGHT

1 ounce = 28 grams

1 gram = 1,000 milligrams (mg)

Scenario #1

A friend gives you 2 ounces (56 grams) of trim that you know is roughly 23 percent THC to make medicine.

First, how much THC total is in the 2 ounces?

56 g × 1,000 = 56,000 mg of plant material

56,000 mg × 0.23 (23%) = 12,880 mg THC

You decide to tincture it in alcohol and do a 1:10 ratio. When you are done, you are left with 560 mL of tincture.

To calculate mg/mL (standard dosage):

12,880 mg THC/560 mL = 23 mg/mL of THC

Now multiply by fudge factor of 80%:

23 mg/mL × 0.8 = 18.4 mg/mL (per 2 droppers)

For 1 dropper (18 drops), that equals:

18.4 mg/1 mL = 9.2 mg/0.5 mL

9.2 mg/18 drops = 0.5 mg/drop

Scenario #2

Instead of making a tincture, you decide to take 2 ounces of an 11% CBD strain and make an infused coconut oil so you can make goo balls for your friend. You will add the plant material to 2 cups of coconut oil. You want to end up making goo balls with 5 mg of CBD per ball. The recipe calls for ¼ cup of coconut oil. How many goo balls can you make?

First, how much CBD is in the 2 ounces?

56 g @ 11% CBD

56,000 mg × 0.11 = 6,160 mg CBD

Now multiply by the 80% fudge factor:

6,160 mg × 0.8 = 4,928 mg CBD

4,928 mg CBD in 2 cups of coconut oil

How much in ¼ cup?

4,928 mg/2 cups = 616 mg/0.25 cup

If I want 5 mg per goo ball:

616 mg/5 mg per goo ball = 123 goo balls

CHAPTER FIVE

CONTRA-INDICATIONS

AND CONSIDERATIONS

As practitioners, every situation we treat is unique. Each person we work with requires a particular dosage and a particular cannabis cultivar. One particular cultivar might work well for a specific condition in one person and not in another. The Western medical model of a standardized dosage for each condition applied across the board for every person is not so useful when it comes to herbalism, and especially with cannabis.

Limiting Factors

Unfortunately, we do not have a lot of data on human studies with cannabis due to our federal laws classifying cannabis as a Schedule I substance. By the U.S. government's definition, cannabis currently has no accepted medical use and a high potential for abuse. (Even though the government owns patents on cannabis and states the benefits of its neuroprotective and antioxidant properties in the patent application.) The best research on cannabis with human subjects is being done outside the United States. Many of the U.S. studies are conducted with pharmaceutical isolates of cannabis. When studies do use flowers, researchers ask subjects to self-report data rather than use the standard protocol of a double-blind study. That said, it is wise to consider any contraindications or potential concerns when using cannabis for medicine.

Addiction

Cannabis is the fourth-most-addictive substance used by Americans; caffeine is number one, alcohol is number two, and tobacco is third. (Frankly, I would rank sugar as number one, but that's a whole other discussion.) Presently, the addiction rate for cannabis is 9 to 10 percent. In comparison, cocaine has a 12 percent addiction rate and alcohol 15 percent. Functional magnetic resonance imaging (fMRI) readings in the brains of addicted chronic cannabis users show that areas of the brain light up in the same patterns as they do for other addictive substances, including alcohol, tobacco, cocaine, heroin, and sugar.

Like other addictive substances, cannabis increases the amount of the neurotransmitter dopamine in the brain. Dopamine is the reward chemical that says, "Do that again!" However, while other addictive substances gradually cause a decrease in dopamine receptors, which creates a tolerance and the need for more of that substance to gain the same pleasurable effect, cannabis does not decrease dopamine receptors. Its mechanism of addiction lies elsewhere.

Risk factors for cannabis addiction include initial use at an early age and use of more potent forms. The quicker and more intensely a substance delivers a high, the higher the risk of addiction. As more people use the recreationally available resin concentrate dabs, with

THC levels as high as 90 percent, we may very well see the cannabis addiction rate rise.

Withdrawal symptoms from cannabis begin one to two days after cessation of use and include irritability, anxiety, decreased appetite, restlessness, sleep disturbance, and sometimes functional impairment. Symptoms usually peak at one week and persist for three to four weeks. After four weeks, cannabinoid receptors return to baseline, and symptoms disappear. It's important to note that if people have been working with cannabis to treat anxiety, going through withdrawal might trigger anxiety, and those people may feel the need to start using cannabis again; they will need to ride out the withdrawal symptom of anxiety.

THE PHYSIOLOGY OF TOLERANCE

When we consume external substances that bind to receptors in our bodies, our bodies try to maintain balance by either decreasing the number of receptors for the substance or decreasing our own chemicals that bind to those receptors (this is termed *downregulation*). Endocannabinoid levels are tightly regulated by the body. If you continuously consume external cannabinoids like THC, the body will say, "Hey, we've got tons of this stuff; let's not make so much" or "Let's make fewer receptors so we aren't so sensitive to all this extra stuff binding to them." When you suddenly stop taking the external cannabinoid, the body can't upregulate the number of receptors or endogenous chemicals fast enough, which causes you to feel withdrawal symptoms. It takes about three weeks for the system to rebalance. Withdrawal from cannabis might be uncomfortable, but it's never lethal.

Allergies

Cannabis as an allergen is, like any plant, specific to an individual. People can develop allergies anytime in life. Windborne cannabis pollen can cause hay fever–like symptoms (this has been documented in the midwestern United States). Touching the plant can cause topical allergies, and inhaling the vapor or smoke can cause respiratory allergies. Hempseed can also cause allergies in sensitive people.

Anxiety

Relaxation and relief from anxiety are two of the most widely reported motives for working with cannabis. Cannabis is generally an effective treatment for anxiety, but there are a few situations where cannabis can cause anxiety. Cultivars with high THC and low CBD can cause anxiety. Most personal-use cannabis has been bred to be high in THC with very little CBD. Some of these cultivars are bred for increased energy and alertness, and can sometimes result in "zippy," anxious feelings. CBD mediates some of these negative effects of THC. As growers and consumers learn more about the plant, cultivars with a more balanced THC/CBD ratio are gaining favor.

Too large a dose of THC can also cause anxiety. You cannot lethally overdose on cannabis the way you can with opioids, but too much THC can mimic the feeling of a panic attack.

One symptom of cannabis withdrawal is increased anxiety. If cessation of cannabis medicine is too abrupt, the body will not have enough time to restart production of endogenous cannabinoids and receptors. Ideally, users should taper use to zero over the course of four weeks to allow the body to increase production of its internal endocannabinoids and receptors.

Cannabis is not meant to take the place of psychotherapy, cognitive or behavioral therapy, or other emotional or psychological work. Cannabis is best used in a supportive role, because one of the plant's great gifts is fostering a sense of safety and well-being.

Cannabinoid Hyperemesis Syndrome (CHS)

Cannabinoid hyperemesis syndrome (CHS) is characterized by cyclic nausea, vomiting, and abdominal pain in chronic cannabis users. The symptoms usually present in the mornings and in individuals under

50 years old who consume cannabis at least once per week for one year or more. Symptoms are sometimes alleviated by a hot bath or shower and abate with cessation of cannabis use. The antiemetic pharmaceutical haloperidol (Haldol) can also alleviate symptoms.

The cause of CHS is unknown, but a top theory is nerve-signaling dysfunction caused by overstimulation of either the CB1 receptor, the TRPV1 receptor, or both. Cannabis allergy is another possible cause.

CHS symptoms can also present in people who are going through withdrawal from cannabis. Additional symptoms of cannabis withdrawal are increased irritability and appetite, sleep disturbances, and depressed mood.

Cognition

During acute cannabis use there is some deficiency in attention, working memory, inhibitory control, and decision-making ability; average IQ drops four to eight points. Long-term users develop a tolerance to some of the negative effects of high-THC strains, such as acute memory deficiency, delayed reaction time, and decreased perceptual motor skills. People do not develop a tolerance to the desirable euphoric feeling.

In studying the long-term effects of cannabis use in adolescents, there is no evidence for a difference in IQ and achievement between users and nonusers when adjustments are made for caregiving environment and tobacco use (tobacco use and environmental conditions play more of a role in IQ than cannabis use). Cognitive deficiencies seen in adults disappear within 25 days of cessation. There are, however, some concerns about brain maturation in chronic cannabis users under the age of 25.

Combination with Other Drugs

Cannabis is anxiolytic, antiemetic, anti-inflammatory, antispasmodic, and anticonvulsant. It supports apoptosis, modulates the immune system, and modulates pain. Any effects of pharmaceuticals used to relieve these symptoms should be monitored; the dose of the pharmaceutical might have to be reduced. Studies show cannabis potentiates the effects of opiates and chemotherapy. Users of cannabis might need lower doses of chemotherapy, opiates, or any other pharmaceutical that affects the same pathways as cannabis.

Contamination

If you are not growing cannabis yourself or do not know the grower, testing for pesticide contamination is crucial. Most commercially grown cannabis is exposed to chemical pesticides, fungicides, and petroleum-based nutrients. Some small-scale farmers also use these products. In the United States, pesticide use is not federally regulated at present, but a few states do have standards. In a recent study, 85 percent of cannabis flowers tested in Colorado were contaminated with pesticides.

Cannabis is a bioaccumulator of heavy metals, and soils where it is grown should be tested. Solvents used for extraction are another serious contamination risk. People who make resin concentrates at home often use butane, a known carcinogen that is not fully purged from the final product. One study found butane levels equal to terpene levels in extracts.

Cultivar Selection

One cultivar of cannabis does not fit all conditions. The practitioner needs to understand the person they are working with and how particular conditions manifest within them. Learning which cultivars are appropriate for specific people and specific conditions is key to effective treatment.

Depletion

In traditional Chinese medicine cannabis is viewed as depleting to chi or vital energy. If cannabis is to be used long term, add a regimen of nourishing food and herbs such as adaptogens (Reishi mushroom, ashwagandha, nettles, and many others) to offset these depleting effects.

Depression

There is no evidence that cannabis users are more at risk of depression than anyone else. The "antimotivational syndrome" of lethargy, apathy, and decreased productivity of cannabis users is a myth. No evidence exists for a causal relationship between cannabis and depression. Depressed individuals may be self-medicating with cannabis.

Detoxification Pathways

As with all new regimens, assessment of the liver's detoxification function is important. Cannabis is eliminated in the liver through the same pathway as most pharmaceuticals. In a healthy liver, cannabis use alongside pharmaceuticals shouldn't be a problem.

Drugs that inhibit that pathway can potentially increase the bioavailability of THC. These drugs include proton pump inhibitors, HIV protease inhibitors, macrolides, azole antifungals, calcium antagonists, and some antidepressants. Drugs that increase the pathway's effectiveness may decrease the bioavailability of THC; these drugs include phenobarbital, phenytoin, troglitazone, and St. John's wort.

CBD is eliminated through different detoxification pathways, the same ones used by the sedative clonazepam (Klonopin). If CBD is dominating these pathways and preventing the liver from removing the drug from circulation, oversedation from clonazepam could result.

Using more than 2 ounces of high-THC cannabis flower per week may increase the anticoagulative effects of warfarin (Coumadin).

Diarrhea

High doses of either THC (up to 1 gram per day of THC) or CBD can cause diarrhea.

Drowsiness and Sedation

Drowsiness or sedation is dependent on dosage and cultivar. Both THC and high doses of CBD can cause sedation. The terpenes myrcene, linalool, and limonene are also sedating.

Dry Mouth and Red Eyes

A dry mouth and bloodshot eyes are symptoms of cannabis use in some people. If the person has a tendency toward dryness, consider introducing moistening herbs such as marshmallow root in conjunction with cannabis.

Gateway Drug

In the United States, cannabis has been called a "gateway drug," meaning it could make a user more likely to try harder drugs, like opioids. There is no causal evidence for this. Rather, evidence suggests that therapeutic cannabis actually decreases rates of tobacco, alcohol, opioid, and prescription drug use and is a viable ally for harm reduction. But you could say that cannabis *is* a gateway drug — into the world of herbalism.

Heart Issues

Both acute and chronic use of cannabis cause low blood pressure. Acute cannabis use may cause tachycardia (increased heart rate) and hypotension (low blood pressure) initially. People at risk of tachycardia or low blood pressure should use caution when beginning a cannabis regimen. Chronic cannabis use can result in bradycardia (a slow heart rate) and hypotension. The switch from tachycardia to bradycardia usually takes 14 days of daily use of high-THC cannabis.

Hypotension and Dizziness

Cannabis use may cause hypotension (low blood pressure). When systemic blood pressure drops, some people are prone to orthostatic hypotension, an inability to increase blood pressure when a person moves from sitting to standing. This short-term inability briefly delays blood flow to the brain, so the person feels dizzy. After a few seconds the body usually readjusts. Orthostatic hypotension may occur with higher THC dosages. Systemic hypotension may result from chronic high-THC cannabis use.

Immune Function

In animal studies, animals that were given 50 to 100 times the psychoactive dose of cannabis showed a decrease in immunity. A psychoactive dose for a new cannabis user might be as low as 2.5 to 5.0 mg. A dose 50 times higher would be over 1 gram; nobody except a person fighting cancer regularly takes doses that high. No evidence exists that cannabis depletes the immune system when used at therapeutic levels. Normal immune systems have been seen in 20-year chronic smokers of cannabis.

Insomnia

Cultivars that are high in THC, low in CBD, and low in myrcene can be too stimulating for individuals and cause sleeplessness. A high-CBD strain doesn't necessarily promote sleep. It can actually be stimulating. In a full-spectrum medicine, some terpenoids are stimulating (terpinolene and pinene) while others are sedating (myrcene, linalool, limonene). You will need to consider the constituents of the particular cultivar.

Nausea

Nausea can be an unwanted effect of a high dose of THC. What is considered a "high dose" varies from person to person, but it is helpful to know this can happen. Animal studies have shown that cannabidiol (CBD) at low dosages decreases vomiting while higher doses of CBD increase it.

Paranoia

One of the gifts of cannabis is to help us open, neurochemically, into a sense of safety. But when some people use cannabis in a vulnerable time or place, they might feel resistance, which can manifest in a sense of paranoia. It is advisable, especially for new cannabis users, to begin with low doses in secure, comforting surroundings.

Pregnancy

While there is documented occurrence of women using cannabis to mitigate pain associated with menstrual and reproductive issues, including migraines, cramps, and leg pain, it is not advisable to use cannabis during pregnancy. Given that the endocannabinoid system is involved in all neural development of a fetus, abstention from cannabis use is advised. When a pregnant woman consumes cannabis, THC and CBD do cross the placenta and reach the fetus (at an albeit low level of 0.8 percent of mom's blood level). Limited use of cannabis for morning sickness or pain management should be weighed against potential risks.

Psychosis

Over the last several decades, cannabis use has increased in the general population while the incidence of psychosis has remained the same. Individuals predisposed to psychosis should avoid cannabis, or at least avoid strains that are high in THC, because THC mimics the effects of the disease by lowering both GABA and glutamate levels, both of which are lower in individuals with psychosis.

Respiratory Issues

Studies show no decrease in lung function or increased risk of lung cancer in 20-year cannabis-only smokers. When cannabis is combined with tobacco, all the risk factors for tobacco appear because of tobacco. Chronic cannabis smokers have a modest risk of bronchitis, but upon cessation or use of a vaporizer the risk diminishes.

Synthetic Cannabinoids

Synthetic cannabinoids (called Spice, K2, and dozens of other names) have made their way into the mainstream. They are a legal blend of various synthetic chemicals. (When the FDA declares one particular chemical in the blend illegal, manufacturers simply alter their formula.) Because synthetics are full agonists of the CB1 receptor, there is an increased risk of adverse effects because they bind more strongly and for longer than THC. The biggest risk is heart attack.

Thinking Outside the Box

When used in the proper setting, cannabis can help us move out of our established patterns of thinking to allow unique opportunities for growth. When approached in a respectful and mindful way, and with a willingness to do our own internal work, we can expect to come away from an experience with cannabis different from how we began the journey.

CHAPTER SIX

CONDITIONS AND CLINICAL APPLICATIONS

As we work with cannabis to address specific conditions in the body, keep in mind that research on the efficacy of cannabis in humans is in its infancy, and scientific research on whole-plant extracts of cannabis is sparse. In this chapter, summaries are offered on research with more attention and merit given to human trials with cannabis, especially the long-term longitudinal studies that have been carried out for hundreds of years by practitioners. We will look at animal or cell culture studies, or studies utilizing artificial isolated constituents, only for clues to where cannabis could be potentially used rather than evidence of its benefit or lack thereof.

There are very few random, controlled trials on the efficacy of whole-plant extracts. There is, however, a wealth of research on isolated constituents of cannabis and pharmaceuticals and the biochemical pathways they interact with. We can glean some potential therapeutic uses for cannabis from this research. For example, we apply information about the effects of serotonin in the body and how serotonin reuptake inhibitors work when using high-CBD cultivars because CBD and CBDA bind to serotonin receptors.

Categories of Conditions

When working with clients and their conditions, it's helpful to categorize common underlying causes of the conditions. When we understand the root of a problem and how to alleviate the cause, we can apply it to other situations with the same root problem. In working with cannabis and the conditions it helps with, we can categorize the conditions in the following ways:

◆ Neurodegenerative diseases: Alzheimer's, Parkinson's, ischemic brain injury

◆ Atypical neurological disorders: autism spectrum disorder, seizure disorders, and epilepsy

◆ Pain

◆ Conditions of the psyche: anxiety, depression, PTSD, schizophrenia

◆ Insomnia and other sleep problems

◆ Nausea and vomiting

◆ Conditions with an endocannabinoid-deficiency component: irritable bowel syndrome, migraine headaches

◆ Conditions of the immune system: cancer, multiple sclerosis

When appropriate, I offer some background information on anatomy and physiology to help you understand how cannabis works with a particular condition (if you just want the top-level facts, you can bypass these sections). I offer a description of each condition; the mechanism of action (some are more in-depth than others); research on effectiveness; and dosage guidelines.

Dosage Guidelines

For all conditions, the starting place for dosage is the minimum effective dose (MED) of a 1:1 CBD:THC oral preparation taken every 4 to 6 hours unless otherwise noted. The art and skill will be in determining what the MED is for each client every time they start a new round of medicine. It is a good practice for users to back dosages down when trying something new because the new medicine might have a different potency. Using the MED is cost-effective, too. Why take a whole dropper full when one drop will get the job done?

I have people start with a drop of tincture (yes, one drop), wait 2 hours to see if they get relief, and then take another drop if they aren't feeling relief. They then wait 4 hours to dose again. This process may seem slow, but that's good, especially for the cannabis-naïve. You do not need to feel high to receive relief, and if you do feel altered at whatever the MED is, tolerance will occur after a few days, and you won't continue to feel high.

Keep in mind that if someone is trying your medicine for the first time after having used isolates or something else that is not a full-plant extract, they will likely find they need a much lower dosage than they are used to, perhaps one-third of the dosage of new medicine versus the old. Remember, some studies found whole-plant extracts to be 330 times more powerful than isolates, so it's wise to start with a one-drop dosage.

In discussing conditions here, I will offer any available evidence based on human or animal studies. For dosage recommendations based on animal studies, I have adjusted for a 150-pound person (the pharmaceutical standard).

Cannabis and Chronic Inflammation

Because chronic inflammation underlies all chronic disease, understanding the physiology of inflammation in general will take us a long way toward understanding how cannabis can help in managing disease and how we can apply this knowledge in our formulas for healing. Plus, biochemistry is fun.

Cannabis works when its constituent chemicals bind to specific receptors on or within cells. In nerve cells, this either increases or decreases neurotransmitter release. In immune cells, this mechanism slows the release of inflammatory cytokines and reduces oxidative stress. The only way our body has to heal any injury is through the process of acute inflammation. (Chronic inflammation contributes to chronic disease, not acute.) The process of acute inflammation allows injured cells to call in their own healing. Acute inflammation brings healing to cells and tissues. The complex cascade of events brings in blood with the needed nutrients, oxygen, and immune cells to help heal. When the body fails to shut down acute inflammation, or when inflammation continues in the absence of triggers, the inflammation becomes chronic. Chronic inflammation causes further damage and is present in most major disease. Reducing chronic inflammation is often the root of addressing chronic diseases.

To successfully do this, we need to know and understand the inflammation process.

The Inflammatory Cascade

The inflammatory process begins when a cell is injured. Injured cells begin making and releasing inflammatory chemicals called cytokines. Cytokines act as chemical flares to attract immune cells to the scene, which clean up debris and kill any pathogens present. The immune cells can also release more cytokines to attract still more immune cells to the party. When the acute situation has been handled, the inflammation should cease.

But not always. Chronic inflammation occurs when immune cells or the chronically damaged tissue cells keep signaling inflammation in the absence of an injury. This becomes a self-perpetuating loop that damages, rather than heals, the tissue. Chronic inflammation can be stopped by stopping the production of inflammatory cytokines. This happens through four different pathways: (1) stopping the work of specific enzymes in the cascade; (2) preventing the release of cytokines from immune cells; (3) decreasing the proliferation of immune cells that make the inflammatory cytokines; or (4) killing the immune cells producing the cytokines (a process known as apoptosis). Cannabis and our endocannabinoid system decrease chronic inflammation by interacting with *every* one of these pathways. Pharmaceuticals work only with the production or release of cytokines.

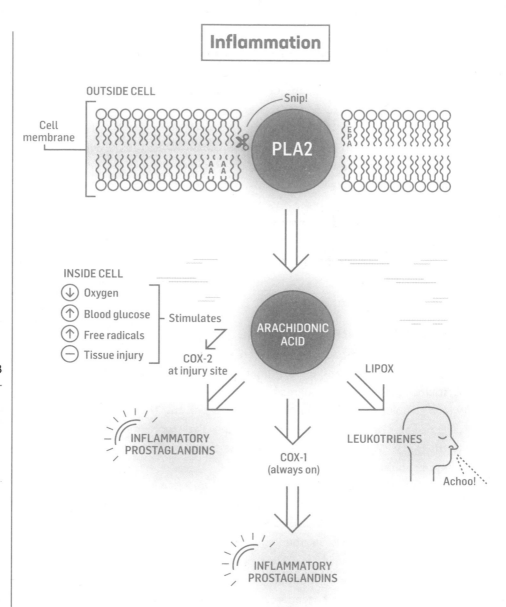

Inflammation

OUTSIDE CELL

Cell membrane

Snip!

PLA2

INSIDE CELL

⊘ Oxygen
⊛ Blood glucose
⊛ Free radicals
⊖ Tissue injury

Stimulates

COX-2
at injury site

ARACHIDONIC ACID

LIPOX

INFLAMMATORY PROSTAGLANDINS

COX-1
(always on)

LEUKOTRIENES

Achoo!

INFLAMMATORY PROSTAGLANDINS

How the ECS and Cannabis Deactivate Inflammation

One of the main jobs of the endocannabinoid system is to shut down inflammation after damage has been repaired. Cannabis and our own endocannabinoids accomplish this using a multipronged approach to stop the release of the inflammatory cytokines. The specific methods are: stopping the production of inflammatory cytokines by activated cells; apoptosis of activated immune cells producing cytokines; and preventing activated immune cells from reproducing so they can't perpetuate the cascade of inflammation.

1. Suppression of cytokine production or stopping the work of specific enzymes in the cascade: The constituents of cannabis and endocannabinoids modulate the inflammatory process at multiple levels (see the diagram on page 172 for specific interventions).
2. Inhibition of cell proliferation: Along with killing the activated cells, cannabinoids bind to CB2 receptors on activated cells, preventing them from making more cells that would continue producing inflammatory cytokines.
3. Apoptosis of activated immune cells producing the inflammatory cytokines: THC binds to CB1 and CB2 receptors on the surface of activated immune cells; T cells, B cells, macrophages, antigen-presenting cells, and dendritic cells (Langerhans cells in the skin) induce apoptosis when these cells produce inflammatory cytokines. CBD activates apoptosis on CD4 T cells (helper Ts) and CD8 (killer Ts), which also overproduce inflammatory cytokines. It's important to note that cannabinoids also protect nontransformed, nonactivated cells from apoptosis; we will discuss this at length in the section on cancer (page 218).

The Pharmaceutical Model

Western allopathic interventions block or shut down one or all of the enzymes involved in the inflammatory cascade. The problem is that the drugs are not specific to the pathways that cause inflammation; they shut down their target enzymes everywhere. Plant medicine does not completely shut down these pathways; they nudge the enzymes rather than push, so the medicine requires ongoing dosing and requires time to produce effects. Pharmaceuticals tend to push rather than nudge, so they are quicker in producing effects, but their lack of specificity has unwanted adverse effects.

Biochemistry of Inflammation

How do cells make inflammatory cytokines? It all starts at the cell membrane. Mechanical, chemical, or physical stimuli at the cell membrane stimulate the enzyme phospholipase A2 (PLA2) within the phospholipid bilayer to snip the tails from the heads of nearby phospholipids. The clipped tails fall into the cell cytoplasm, where they interact with the next set of enzymes.

The enzyme cyclooxygenase 1 (COX-1), termed the "house-keeper," is always active within cells. COX-1 converts omega-6 fatty acids like arachidonic acid into inflammatory cytokines such as interleukin-1 (IL-1), interleukin-12 (IL-12), interleukin-18 (IL-18), and tumor necrosis factor alpha (TNF-A). COX-1 can also enzymatically cleave arachidonic acid to create inflammatory prostaglandins. Prostaglandins are chemicals that act more like hormones to increase or decrease inflammation (hormones travel farther than cytokines to affect cells farther away).

But the same enzyme COX-1 converts omega-3 fatty acids like eicosapentaenoic acid into anti-inflammatory prostaglandins and anti-inflammatory cytokines. The body can start up or shut down inflammation at any time depending on which fatty acids are available.

The standard American diet has a ratio of 20 to 30 omega-6 fatty acids for every omega-3. That 25:1 ratio sets us up to over-produce inflammatory cytokines because of the components we give to the enzymes. Increasing the amount of omega-3 fatty acids in the diet can nudge cell membranes back into balance, with one omega-3 for every omega-6.

This is not a quick fix. Research has shown that it may take up to eight months to bring more balance even when taking 1,500 mg of omega-3 fatty acid supplements a day. You have 50 trillion cells with millions of fatty acid tails in each cell. Replacement takes a long time, and the exact 1:1 fatty acid ratio probably will never be attained. But we can shift back into balance.

The enzyme cyclooxygenase 2 (COX-2) is also within the cell, but activates at sites of inflammation and is more affected by oxidative factors. Decreased oxygen, increased blood glucose, oxidative stress, and an increase in free radicals all stimulate COX-2. COX-2 has the ability, like COX-1, to make both pro-inflammatory cytokines and prostaglandins and anti-inflammatory cytokines and prostaglandins, depending on the type of fatty acids available.

Arachidonate 5-lipoxygenase (5-LOX) is another intracellular enzyme that interacts with the cleaved phospholipids. Action of this enzyme on arachidonic acid produces a different class of cytokines, the leukotrienes. Leukotrienes induce all the symptoms of allergy and can be enzymatically broken down further into tumor necrosis factor and interleukin-1, two more pro-inflammatory cytokines.

After these cytokines are made, they are released from cells, attracting more immune cells, increasing blood flow to the area, and increasing capillary permeability, allowing nutrients, oxygen, and immune cells to migrate more easily to the injured area.

Once the acute situation has been resolved, the inflammation cascade stops. Until recently, inflammation cessation was thought to occur when the inflammatory cytokines were removed from the area. Recent evidence shows two other mechanisms for the cessation of inflammation and another contributing factor to chronic inflammation. First, the breakdown of arachidonic acid and eicosapentaenoic acid by enzymes can be switched from making inflammatory cytokines to making anti-inflammatory cytokines: resolvins, lipoxins, and protectins. The mechanism behind this switch is unknown. The second mechanism of cessation of inflammation is brought about by, you guessed it, the endocannabinoids.

CONDITIONS AND CLINICAL APPLICATIONS

Biochemistry of Inflammation continues on next page

Inflammation Intervention

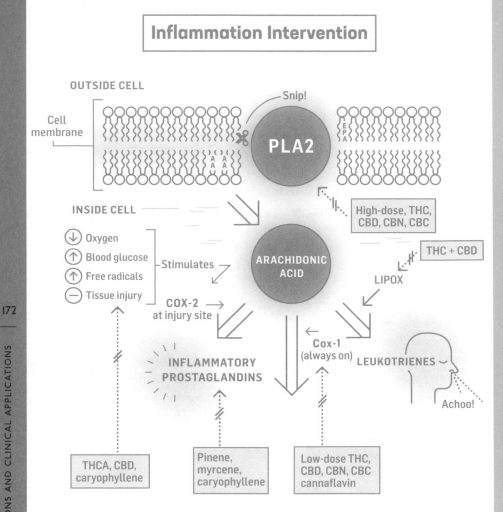

Neurodegenerative Diseases

All diseases with a neurodegenerative component, including Alzheimer's, Parkinson's, and Huntington's, share an initiating factor: neuroinflammation, or inflammation of neural cells. Luckily, we can apply our understanding of chronic inflammation and the ways cannabis ameliorates it to the nervous system. Neuroinflammation involves some additional cells: microglia and astrocytes, considered the immune cells of the nervous system.

The Mechanism of Neuroinflammation

1. Neuronal cell injury stimulates the release of inflammatory chemicals.
2. Microglia migrate to the injured area to help with healing, where they become activated and begin secreting inflammatory cytokines and making free radicals.
3. Inflammatory cytokines stimulate the migration of immune cells to the area, which remove debris and fight infection. In the case of the neurodegenerative diseases, inflammation continues and becomes chronic.
4. Activated microglia continue secreting inflammatory cytokines and recruiting more microglia, which exacerbates local cell damage.
5. Activated microglia continue producing free radicals, causing further damage.
6. Activated microglia increase the release of glutamate, causing further excitotoxicity and damage to neurons.
7. As inflammation progresses, astrocytes, which form the blood-brain barrier, become compromised, and the blood-brain barrier becomes partially disrupted. This allows more immune cells (macrophages, natural killer cells, and antibodies, which are not normally found in the brain) from the blood to migrate to the site of injury and start producing yet more inflammatory cytokines.
8. The continued inflammation, excitotoxicity, and production of free radicals leads to decreased neuronal function and, ultimately, neuronal cell death, which leads to decreased brain function.

Neural Protection

When we talk about "neural protection," we mean preventing damage to neurons, which for the most part do not regenerate readily. Our first line of neural protection lies in the natural process of shutting down inflammation by secreting anti-inflammatory cytokines and endocannabinoids. When inflammation begins, immune cells upregulate their CB2 receptors, which leads to increased binding to CB2 receptors by endocannabinoids. The net effect is a decrease of inflammatory cytokine release by the activated cells. As the disease progresses, the immune cells that have crossed the blood-brain barrier also upregulate their CB2 receptors, making them more susceptible to CB2 agonists. Cannabis has four neuroprotective mechanisms:

1. It is anti-inflammatory and stimulates the production of anti-inflammatory cytokines.
2. It decreases the production of inflammatory cytokines.
3. It is an antioxidant and helps protect neurons and neuroglia from the damaging effects of free radicals.
4. It decreases excitotoxicity by decreasing the release of the excitatory neurotransmitter glutamate.

Cannabis as an Antioxidant

The beauty of cannabis as an antioxidant is that (similar to its benefits as an anti-inflammatory agent) it interfaces at multiple sites. It is more potent than vitamins A and E as an antioxidant and decreases reactive oxygen species, lipid peroxidation, and nitric oxide (free radical) production during acute inflammation.

Oxidation and Antioxidants

We've all heard about free radicals and how they contribute to disease and are not good for us. But what are they exactly and why are they so dangerous? How do antioxidants play into the story? And why are we discussing them here in conjunction with neurodegenerative diseases?

Throughout the day as we go about our life we are making energy in the form of adenosine triphosphate (ATP) in our mitochondria. We are fighting off foreign cells and probably recovering from some sort of injury. All of these reactions create free radicals. They are necessary and in some cases even helpful. These processes combine oxygen with another element, a process called *oxidation*. When oxygen is combined with another element, electrons are lost and are free to move around. Here we have the free radicals, or reactive oxygen species (ROS), which can be damaging to structures nearby. Specific ROSs you may have heard of include hydrogen peroxide, superoxide dismutase, hydroxyl radical, and singlet oxygen. Oxidation occurs continously in our bodies and is often paired with a reaction where an electron is gained, termed "reduction."

If free radicals are uncoupled (unpaired) they cause damage to fragile structures within the cell. Think of them as bouncy balls thrown inside a room with glass figurines. The glass figurines, if hit just right with the bouncy ball, will be broken. Fragile structures at risk of being damaged are DNA, RNA, the polyunsaturated fats that make up your cell membranes (arachidonic acid — remember that one?), and enzymes within cells. The brain cells are especially sensitive to destruction because of their high lipid content. Further, damage to the nervous system is more serious because the cells do not regenerate as easily as other cells.

Oxidation is not only a destructive process. Our bodies use this highly reactive process to maintain health. Where? In cellular respiration within the mitochondria of every one of your cells.

Oxidation and Antioxidants continues on next page

175

We literally burn glucose with oxygen in our powerhouses, the mitochondria. Our immune cells use oxidative by-products to kill pathogens. Platelets use them for wound repair, the liver uses it to detoxify everything we come into contact with, and oxidation signals the natural cell death of apoptosis.

Internal factors that increase oxidation are mitochondria increasing production of ATP (our cells need more energy), ischemic injury (during stroke or heart attack), increased immune response, and aging. External factors that increase the production of free radicals are pollutants, tobacco, smoke, drugs, radiation, and xenobiotics.

The unpredictable and uncontrolled nature of the production of free radicals can lead to imbalanced states and unwanted apoptosis of cells, necrosis of tissue, and cell death. Conditions with oxidative stress as a component include neurodegenerative diseases (Huntington's, Parkinson's, Alzheimer's, stroke, amyotrophic lateral sclerosis, and multiple sclerosis); diabetes mellitus; rheumatoid arthritis; ischemia reperfusion injury found in cardiovascular disease, stroke, and brain injury; hypertension; obesity; metabolic syndrome; atherosclerosis; irritable bowel syndrome; depression; autism; cancer; ADHD; cataracts; macular degeneration; inflammation; and emphysema.

The balance of health within us rests in our ability to repair and detoxify the effects of oxidation on a daily basis. Fortunately, we have many systems of repair in place within us. A healthy, balanced body is a system where damage from oxidation is balanced by antioxidant defense.

The beauty comes with the understanding that we also hold within us processes for maintaining balance and health. We have substances for capturing the bouncy balls bouncing around. They are the antioxidants. We experience disease when we are out of balance. This state of too many oxidative by-products and not enough antioxidants is termed oxidative stress.

The number one antioxidant within the blood, uric acid, doesn't get a lot of press. Other water-soluble antioxidants we're carrying around are vitamin C and glutathione. Lipid-soluble antioxidants are vitamins E and A. Enzymes that remove free radicals are the peroxidases and superoxide dismutases.

There's more help. Our ancestors, the plants, developed and evolved in a potentially lethal and toxically high-oxygen environment, where the production of free radicals was commonplace. The solution the plants came up with was a catcher's mitt of sorts to capture the damaging free radicals. Enter the antioxidants. Antioxidants "catch" and remove the free radicals and prevent other oxidative reactions. We obtain the miraculous antioxidants by simply eating whole plants with their antioxidants inside. Humans benefit from eating plants and absorbing these antioxidant molecules!

Lifestyle choices we make can also increase antioxidants: exercise; cessation of smoking; decreasing chemical toxin exposure; decreasing the damaging effects of the sun; sleep; decreasing consumption of alcohol, sugar, and unhealthy fats; eating a rainbow of fresh plants; and, lastly, eating enough to sustain our life rather than overeating.

Alzheimer's Disease

Alzheimer's is the most common form of dementia. It is a progressive neurological disorder characterized by the production of plaques and neurofibrillary tangles in the brain. The plaques are made of amyloid beta peptides, which block neurons from firing, trigger inflammation, and activate microglia. The protein tangles, caused by oxidative stress, block the transport of nutrients and other essential ingredients for neuronal function. The tangles also prevent microglia from moving in and cleaning up debris. Together this leads to the progressive destruction of synapses, neuronal cell death, memory loss, and decreased cognitive function.

Symptoms of Alzheimer's that may respond to cannabis include loss of sleep, paranoia, anxiety, pain, behavioral disturbances, poor appetite, and weight loss.

Mechanism of Action

Cannabis is neuroprotective, decreases inflammation, neutralizes reactive oxygen species (ROS), and reverses excitotoxicity. These actions mitigate the damage done by inflammation and free radicals to neurons, increasing synaptic transmission.

Research on Effectiveness

Rodent studies of early-onset Alzheimer's have found that preparations of 1:1 THC:CBD decrease inflammation brought on by activated microglia, decrease beta amyloid deposits and inflammatory markers, and improve memory deficits.

CBD-only preparations increased cell survival of embryological neuronal cells, decreased production of ROS, nitric oxide, and neurofibrillary tangles, reduced inflammation, and activated macrophages to remove beta amyloid plaques.

THC-only preparations prevented beta amyloid–induced neurotoxicity and helped reverse memory deficits and cognitive dysfunction. THC binding to CB1 receptors also inhibits production of acetylcholinesterase, which enhances synaptic transmission and stimulates neurogenesis, both of which ameliorate cognitive dysfunction.

THC binding to CB2 receptors reduces inflammation and activates macrophages to remove beta amyloid plaques.

Human studies show the addition of THC to existing regimens significantly improved behavior disturbances such as delusions, agitation, irritability, apathy, sleep, food intake, and overall symptom severity.

Dosage
Minimum effective dose (MED) for alleviation of symptoms.

Parkinson's Disease

Parkinson's disease, the second-most-common neurodegenerative disease, is characterized by a progressive degeneration of dopaminergic (dopamine-producing) neurons in the midbrain, resulting in severe motor impairment and loss of motor control. Current pharmacological therapies consist of administering L-dopa, which gives some relief initially but decreases in effectiveness over time.

Mechanism of Action
Because cannabis offers neuroprotection and lowers oxidative stress, it may also slow the overproduction of ROS in neurons trying to upregulate their production of dopamine. If these neurons experience less damage, they should be able to produce dopamine for longer. Microglial activation may also be attenuated by cannabis's ability to decrease inflammation.

The ECS regulates dopamine production via the CB1 receptors. Dopamine release is regulated by GABA and glutamate. Both of these neurotransmitters are regulated by endocannabinoids (or phytocannabinoids) binding to the CB1 receptor. Acute THC use increases dopamine release, while chronic THC use decreases dopamine release. CBD is also a partial agonist of the dopamine receptor.

DEEP DIVE

Cannabis and Neuroprotection

THC binds to CB1 and CB2 receptors, and caryophyllene to CB2 receptors, on immune and microglia cells to cause:

- ◆ Decreased inflammatory cytokine production

- ◆ Increased anti-inflammatory cytokine production

- ◆ Decreased adaptive immunity (lower B and T cell production and function)

- ◆ Decreased proliferation and maturation of activated immune cells

- ◆ Induction of apoptosis of CD4 T cells that destroy the blood-brain barrier

THC binds to CB1 receptors to:

- ◆ Inhibit glutamate-driven excitotoxicity and decrease nitric oxide (NO) production

- ◆ Inhibit astrocyte production of inflammatory cytokines and NO and increase the production of IL-6 (which increases neuronal growth factor and decreases production of TNF alpha)

- ◆ Decrease cellular respiration in the mitochondria, which leads to decreased ROS and free radicals

CBD in cannabis increases AEA, which binds to CB1 receptors and does all the above listed for THC.

Research on Effectiveness

Human studies with 75 mg or 300 mg of CBD per day showed no motor improvement but did improve quality-of-life scores. Dosages of greater than 150 mg of CBD alone did show improvement of psychotic scores, but no motor improvement. Israeli researchers found inhaled cannabinoids improved motor symptoms of tremors, rigidity, and slowness of movement, and a 30 percent decrease in pain intensity, for people with Parkinson's.

Dosage

Begin with 25 mg of a high-CBD strain over the day and move up to 150 mg if needed. At 50 mg of CBD, begin doses of THC starting at 1 mg per mL.

Ischemic Brain Injury

Hypoxic damage to the brain from stroke, cardiac arrest, near drowning, birthing complications, or trauma can result in immediate and progressive cognitive decline.

Mechanism of Action

Endocannabinoids and phytocannabinoids both bind to CB1 and NMDA (glutamate) receptors, offering protection against excitotoxicity from excess signaling by glutamate and oxidative stress. The body increases 2AG naturally after injury as a protective mechanism. It binds to CB1 receptors to further help attenuate the excitotoxicity.

Research on Effectiveness

Research with mice showed increases in post-injury recognition memory and spatial memory and decreased apoptosis of neurons and decreased inflammation with CBD dose equivalents of 68 mg to 2,040 mg both pre- and post-injury.

Dosage

Start at 15 mg CBD two to three times per day and work up to a total of 500 mg per day.

QUALITY OF LIFE

Cannabis has been documented in study after study as improving quality of life for people suffering from chronic conditions. This benefit is often dismissed or noted with an eye roll. But cannabis should be added to formulations *because* it increases quality of life. If you suffer from chronic pain, your quality of life is diminished. If cannabis gets you back to enjoying life more, you will do more things that bring you joy. Maybe you will feel well enough to play with your grandchildren, or garden, or go for a walk. These things also improve your quality of life. And guess what? When you experience improved quality of life, you make more of your own painkillers and mood enhancers: endorphins. You also make more endocannabinoids to fight inflammation and cause bliss. It becomes a positive feedback loop of goodness!

Atypical Neurological Conditions

Some people with atypical neurological conditions (including autism spectrum disorder, seizures, and epilepsy) have been shown to respond positively to cannabis.

Autism Spectrum Disorder (ASD)

Autism spectrum disorder (ASD) affects 1 percent of the U.S. population; 10 to 30 percent of those individuals have concurrent epilepsy. Forty percent of people with ASD do not respond to standard pharmaceutical treatment. Current pharmaceuticals used — many of them

182

CONDITIONS AND CLINICAL APPLICATIONS

with children — include antipsychotics, mood stabilizers, benzodiazepines, serotonin reuptake inhibitors, and stimulants. We've recently
begun to question the ethics of giving strong psychoactive drugs
to children; especially since the drugs have not been tested for that
population.

Children and adults with ASD are deficient in social-reward processing and have atypical responses to facial expressions and emotions. Babies who neurologically feel less reward for gazing into the
happy faces of their caregivers will look less often and have fewer
social interactions than neurotypical children.

Mechanism of Action

The ECS oversees cognitive function, emotional regulation, social
function, motivation, and reward processing. A diminished ECS is a
possible mechanism of action for ASD.

Imaging has shown that ASD people have abnormalities in volume, connectivity, and activity in the reward centers of the brain that
respond to social stimulus. The phasic release of dopamine in the
reward centers is the primary mechanism encoding reward behaviors.
Dopamine release is regulated by GABA and glutamate. Both of these
neurotransmitters are regulated by endocannabinoids binding to CB1
receptors. This is a possible link between the ECS and ASD. ASD people have fewer CB1 receptors and lower AEA blood levels. AEA levels
are also lower in people with temporal lobe epilepsy.

Research on Effectiveness

CBD-rich cannabis is known for its anxiolytic, antipsychotic, and immu-
nomodulatory effects and for increasing endogenous AEA. Rodent and
human studies show the benefits of CBD-rich cannabis for ASD.

In one human study, adults took 68 to 720 mg of CBD per day
(using a CBD:THC ratio of 20:1) for 7 to 13 months; 61 percent showed
improved behavior, 39 percent showed improved anxiety, and 47 percent showed increased communication. Thirty-three percent of the
participants used fewer meds or decreased their dosage, 24 percent
stopped meds altogether, and 8 percent used more. The authors of
the study recognized that one flaw was that not all participants used
the same CBD-rich cannabis.

Dosage

Based on the few human studies, start at 50 mg of 20:1 CBD:THC per day and work up as needed.

Seizure Disorders and Epilepsy

Anti-epilepsy drugs are the leading treatment for childhood epileptic seizures. One-third of people with epilepsy do not respond to two or more of these drugs and are deemed intractable or drug resistant. Twenty new seizure medications have been developed in the past decades without much improvement, so it's understandable that parents are looking for alternatives.

A young girl named Charlotte Figi, who has a rare seizure disorder called Dravet's that caused her to experience 300 seizures per week, and her mother, Paige Figi, breathed life into the paradigm shift we are seeing regarding the use of cannabis as a treatment option for seizures. It's been a long time coming — cannabis was documented on Sumerian Akkadian tablets in 1800 BCE as being used for "night convulsions."

Mechanism of Action

The precise cause of epileptic seizures is unknown, but they involve an electrical storm of neural activity within the brain. This electrical activity is caused by an increased release of the excitatory neurotransmitter glutamate leading to excitotoxicity and behavioral deficits.

PHYSIOLOGY OF A SEIZURE

When neuronal activity (as seen in seizures) increases, endocannabinoids are produced on demand to bind to presynaptic CB1 receptors, decreasing the release of glutamate and increasing the release of GABA and therefore decreasing seizures. Over the long term, CB1 receptors are upregulated. Cannabis modulates excitation in a multitude of ways:

- THC binding to CB1 receptors decreases glutamate release and increases GABA release. (A net effect of decreasing neuronal excitement.)

- THC is anticonvulsant at subsedation levels.

- THCA, CBDV, THCV, and the terpene linalool are all also anticonvulsant.

- CBC is anticonvulsant and influences adult neural stem cell differentiation by slowing the generation of new astrocytes involved in neuroinflammation.

- CBDA is anticonvulsant and also has 100 times higher affinity for the 5-HTA (serotonin) receptor than CBD.

- CBD is a superstar anticonvulsant due to its multipronged approach. CBD increases inhibitory transmissions and decreases excitatory transmissions of the neurons via:

  - Increasing GABA receptor inhibition and calcium channel movement (both anticonvulsant)

  - Positive allosteric modulation of GABA, which increases the inhibitory effect of nerve transmission by GABA (leading to decreases in excitatory nerve transmission)

  - Decreasing the presynaptic release of glutamate by binding to the cation channel on the presynaptic neuron (leading to decreases in excitatory nerve transmission)

  - Decreasing adenosine reuptake, which inhibits synaptic transmission

  - Indirectly activating adenosine

  - Activating 5-HTA (serotonin) receptors

  - Antioxidizing

  - Acting as an anti-inflammatory

Research on Effectiveness

Dosing ranged from 0.02 mg/kg (1.4 mg for a 150-pound person) to 50 mg/kg (3,400 mg for a 150-pound person). The lower range was always used with whole-plant extract, and the higher range was used for purified CBD extracts.

The generalized results of 10 studies on 1,400 people are as follows:

◆ Fifty-one percent of participants saw a reduction of seizures by at least 50 percent.

◆ Sixteen percent of participants became seizure-free.

◆ An increased quality of life, including better mood, increased alertness, improved sleep, positive behavior changes, and increased language and motor skills, were seen in those who used whole-plant extracts but not in those who used purified extracts.

◆ Negative effects included appetite changes, drowsiness, gastro-intestinal disturbances, weight changes, fatigue, and nausea.

◆ Severe adverse effects are rare and happen when people are taking CBD with other anti-epilepsy drugs. They include thrombo-cytopenia, respiratory infections, and altered liver enzymes.

◆ CBD is at least as safe as other anti-epilepsy drugs, if not more so.

◆ CBD-enriched (whole-plant) extracts are more potent and have a better safety profile than other anti-epilepsy drugs.

◆ A four-year study of the treatment-resistant population consisting of 670 people found that whole-plant extracts were four times as potent as purified extracts. The study also showed that whole-plant extracts caused a 71 percent reduction in seizures, whereas the purified extracts produced a 36 percent reduction. That means that whole-plant dosages are one-fourth the dosage of purified extracts, resulting in less cost.

Dosage

We must remember again to use the minimum effective dose. Reduction of seizures has been found at just 0.02 mg/kg. That's

1.4 mg for a 150-pound person (pharmaceutical companies recommend a starting dosage at 10 times this amount!).

Pain

Pain relief is the most commonly cited reason for cannabis use and is our first known record of use. It is one of three conditions fully endorsed by the National Academies of Sciences, Engineering, and Medicine. To fully understand the multilayered brilliance of cannabis's interaction in the body to relieve chronic pain, we will take an in-depth look at the pain pathway itself and then look at the mechanisms through which cannabis intervenes.

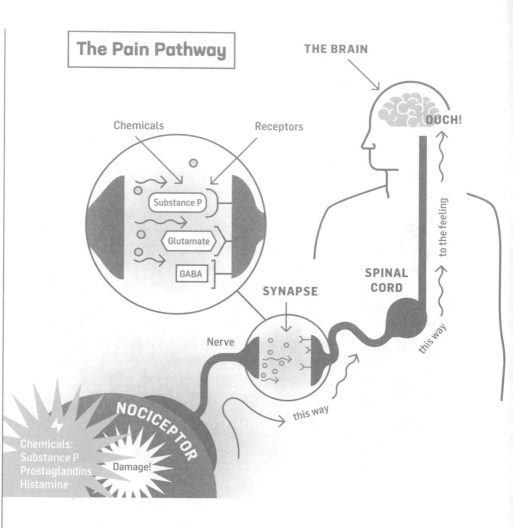

The Pain Pathway

THE BRAIN

OUCH!

Chemicals

Receptors

Substance P

Glutamate

GABA

SYNAPSE

SPINAL CORD

to the feeling

Nerve

this way

this way

NOCICEPTOR

Chemicals:
Substance P
Prostaglandins
Histamine

Damage!

Mechanism of Action

Chronic pain differs from acute pain in that it lasts longer (at least three months) and is maladaptive (doesn't shut off automatically). Four categories of chronic pain exist: nociceptive, inflammatory, neuropathic, and dysfunctional. We will discuss nociceptive, inflammatory, and neuropathic.

In nociceptive and inflammatory pain, damaged cells release chemicals that initiate inflammation and healing. Substance P (SP) is one such chemical; it binds to NK-1 receptors on sensory neurons carrying pain signals, triggering electrical nerve impulses that carry the pain message up to the brain.

Into the Synapse

Amazingly, cannabis works at each one of the following sites to alleviate pain.

Substance P and the NK-1 Receptor

Two chief actors in the pain mechanism are the neurotransmitter substance P (SP) and its receptor, neurokinin 1 (NK-1). Along with signaling pain, SP is also involved with vasodilation, inflammation, regulation of nausea and emesis, regulation of mood, anxiety, stress, neurotoxic pain, neurogenesis, cell growth, and wound healing. Substance P is the main excitatory neurotransmitter in the entire pain pathway from nociceptor all the way up through the brain. THC decreases SP. One noninvasive way to decrease SP is to sleep.

Endorphins and Opioid Receptors

One circuit breaker we have for relieving pain is beta endorphin. When released, endorphins bind to their three opiate-receptor sub-types, mu, kappa, and delta, on presynaptic neurons along the pain pathway. This prevents the release of SP, halting the nerve impulse and relieving pain. Endorphins are natural pain relievers; anything that brings you joy triggers a release of endorphins. Pharmaceutical opiates (exogenous endorphins, if you will) bind to mu receptors and also decrease the release of SP. Cannabis binds to all three of the opiate receptor subtypes to decrease SP.

The pathway is not one long, unbroken cord but rather many neurons in a path, separated by synaptic gaps. The nerve impulse cannot jump across these gaps. Instead, chemical messengers (neurotransmitters) are released that cross the gap and bind to receptors on the next neuron, which carries the electric impulse along. Synapses for the pain pathway occur in the spinal cord, brain stem, thalamus, limbic system, and sensory cortex of the cerebrum. Neurotransmitters include substance P and glutamate (both of which are excitatory and keep the impulse going) and GABA, which is inhibitory (it stops the signal). The end point for pain signals is the sensory cortex in the brain. Once the signal arrives there, you become conscious of the pain and know where in the body it is. As you already know from experience, it takes less than a second to stub your pinky toe and feel the pain.

Endocannabinoid System and Pain

With this understanding of the pain pathway, let's look at the endocannabinoid system. Endocannabinoids are produced, on demand, in both neural and nonneuronal cells in response to tissue injury and excessive pain signaling. They suppress inflammation and decrease sensitivity to pain.

CB1 AGONISTS (AEA, 2AG, THC, CBN)

The endocannabinoids AEA and 2AG, as well as THC and CBN, mediate pain by binding to CB1 receptors, which decreases the transmission of the pain impulse by decreasing the release of SP and glutamate. They also increase the production of endorphins.

TRPV1 AGONISTS (AEA, 2AG, PEA, OEA, HIGH-DOSE CBD, CBN, CBG, THCV, CBCA)

CB1 receptors exist on presynaptic neurons within the central nervous system, colocated with vanilloid (TRPV1) receptors. Vanilloid receptors, when bound to by some agonists, trigger the transmission of pain signals. All the agonists listed above bind to and desensitize vanilloid receptors, making them less sensitive to agonists and less apt to transmit a pain signal.

CB2 AGONISTS (AEA, 2AG, THC, CARYOPHYLLENE)

CB2 receptors are found on immune cells, some neurons, and the microglia. Baseline levels of CB2 receptors on microglia are low, but under pathological conditions CB2 receptor production is upregulated. When 2AG, AEA, THC, or caryophyllene binds to CB2 receptors, the release of inflammatory cytokines slows and neurons are protected from oxidative damage.

CBD boosts the effects of AEA by preventing its breakdown by FAAH. (Nonsteroidal anti-inflammatory drugs [NSAIDs] work this way too.) CBD also potentiates the analgesic effects of THC. In the skin, endocannabinoids binding to CB2 receptors on keratinocytes (skin cells) stimulates the release of beta endorphin, which binds to mu receptors and decreases pain.

BOOSTING ENDORPHINS

- Do what you love!

- Acupuncture

- Exercise boosts endocan-nabinoids as well. Both are the neurotransmitters behind "runner's high."

- Massage also increases endocannabinoids

- Good sex

- Myrcene — one of the terpenes in cannabis

Microglia Function in Pain

Microglia contribute to chronic pain via two main mechanisms. First, when trauma occurs in the nervous system, microglia activate and secrete inflammatory cytokines to increase healing and the immune response, causing further inflammation and pain. Activated microglia then increase the number of CB2 receptors on their surface and increase their production of AEA, decreasing pain and inflammation.

This mechanism of activated microglia becoming more sensitive to agonists binding to CB2 receptors, and decreasing the release of inflammatory cytokines and relieving pain, initiates the cessation of the inflammatory response.

Neuropathic pain found in people with cancer, diabetes mellitus, multiple sclerosis, and peripheral nerve injuries can be alleviated by CBD's inhibition of microglia activation and migration within the spinal cord and brain.

The second mechanism is carried out by a separate receptor on the microglia, the TL4-R (toll-like receptor). When the TL4-R is bound to by a class of chemicals called alarmins, immune-stimulating and neuroexcitatory cytokines (IL-1, IL-6, and TNF) are released. Studies have shown that blocking the release of these cytokines decreases pain. Where do the alarmins come from? When *any* cells of the body are damaged (from inflammation, physical trauma, or chemotherapy) the damaged cells release alarmins into the blood; they eventually end up in the cerebrospinal fluid of the central nervous system. From here they diffuse out to interact with the TL4-R receptors on the microglia, signaling pain.

OPIATE-RECEPTOR AGONISTS
Opioid and cannabinoid receptors are both present in pain-signaling regions in the brain and spinal cord, and their pathways interact.

MORPHINE AND INCREASED PAIN PERCEPTION

Sometimes morphine can cause *more* pain. It does this by binding to TL4-R receptors on microglia, like alarmins, releasing more excitatory cytokines and signaling more pain.

Administering cannabinoids (as low as 1 mg THC) with opioids results in a greater-than-additive analgesic effect; THC can potentiate opiates by 30 percent.

INFLAMMATORY PAIN

The prostaglandins that initiate inflammation can also initiate the pain response. AEA, 2AG, THC, THCA, CBD, CBC, myrcene, pinene, and caryophyllene all slow the release of inflammatory prostaglandins. Studies have found that cannabis helps with inflammatory pain due to burns, fibromyalgia, irritable bowel syndrome, atopic dermatitis, and pancreas and pelvic pain.

MUSCLE-SPASM PAIN

Chronic or unwanted spastic contraction is another cause of chronic pain. Neurons in the brain initiate a nerve impulse that travels to a synapse at the muscle, where the neurotransmitter, acetylcholine, crosses the synaptic gap, binds to its receptor, and signals the muscle to contract. Spastic pain occurs when the signal continues to fire when not initiated by the brain. We work with this type of pain using antispasmodics and sedatives.

Antispasmodics decrease chronic contraction by decreasing excitatory-neurotransmitter release at the synapse of the neuromuscular junction. THC, CBD, and myrcene work here. Their antispasmodic properties relieve pain related to chronic contraction and multiple sclerosis spasticity.

Sedatives work within the central nervous system by decreasing the excitability of the spinal motor neurons that stimulate muscles. They work on the whole system rather than targeting just motor neurons, so they make people feel sleepy or sedated. Sedatives are also used as anti-anxiety medications and as anticonvulsants. Benzodiazepines, Jamaican dogwood, California poppy, THC and myrcene in cannabis and hops, all increase the release of the inhibitory neurotransmitter GABA to cause sedation.

Research on Effectiveness

The National Academies of Sciences, Engineering, and Medicine in 2017 released its report indicating "conclusive high-quality evidence" (its highest rating) for the use of cannabis for chronic neuropathic pain and cancer pain.

The Canadian Pharmacists Association in 2018 released an evidence guide for the medical use of cannabis that indicated "moderate quality evidence" for cannabis with general or acute neuropathic and cancer pain.

A 2015 review of the medical use of cannabis for chronic pain and cancer pain looked at 28 studies totaling 2,454 people. The review found moderate-quality evidence for cannabis as a pain medicine. Smoked cannabis provided the highest reduction of pain (probably due to its whole-plant nature).

In a recent U.S. study of 2,400 people, 80 percent of respondents found CBD to be very effective. More than 66 percent of respondents reported that CBD was more effective than their prescribed medications, and 42 percent reported replacing their prescribed medication with CBD. Fifty-four percent of the respondents used CBD for joint pain, 35 percent used CBD for muscle tension with cluster headaches, and 32 percent used CBD for other forms of chronic pain.

A 2016 seven-month Israeli study of 176 people with chronic pain looked at cannabis, opiate use, and quality of life. Individuals were able to dose as frequently and as high a cannabis dose as they deemed necessary. Forty-four percent stopped opiate use, 39 percent decreased opiate dosing, 80 percent had an increase in function, and 87 percent reported an increase in quality of life!

A 2018 review by the National Cancer Institute's Cancer Integrative, Alternative, and Complementary Therapies Editorial Board stated that cannabis contributes to pain relief for people living with cancer.

Finally, a summary from a recent review of 20 clinical trials states that using whole-plant extracts has been shown conclusively to provide more effective pain control than THC alone.

TYPES OF PAIN CANNABIS WORKS WELL FOR . . . SO FAR

Cannabis works well for spinal cord injury, peripheral neuropathy, nerve injury, brachial plexus damage, amputation, phantom pain, rheumatoid arthritis, complex regional pain syndrome, musculoskeletal pain, fibromyalgia, migraine, and cancer.

It is not recommended for acute pain.

Dosage

Pain-management doses start at a 1 mg oral dose of 1:1 THC:CBD every 4 to 6 hours, working up to 10 to 15 mg per dose if needed. Evidence suggests that increasing the dose above 15 mg does not increase pain relief. Use inhalation methods for breakthrough pain.

Cannabis can provide relief from pain either alone or in combination with opioids. Oral dosing for use in combination with opiates starts as low as 1 mg THC per opiate dose.

OPIATES

Opiates are one of three major classes of pharmaceuticals used for pain management. They are not very effective for chronic pain (hyperalgesia occurs with chronic use), and they are dangerous, with significant adverse effects: 1 in 33 users becomes addicted or dies. Fifty percent of people who take opiates longer than 30 days have been found to still be using them at three years. Eighty percent of heroin users begin with prescription opiates. Ninety people die per day from opioid overdose in the United States.

A report from Medicare states that on average, 23 million doses of opioids were used per day every year from 2010 to 2015. States with legal medical cannabis saw an average reduction of 3.742 million daily doses of opiates due to dispensary sales and a reduction of 1.792 million daily doses due to home cultivation of cannabis.

Conditions of the Psyche

Before addressing conditions of the psyche — anxiety, depression, schizophrenia, and posttraumatic stress — it's helpful to delve into some basic brain physiology and the role of the ECS in stress response, because all of these conditions share some common physiology. Understanding this shared physiology will help us understand how cannabis works with each condition.

The ECS and the Stress Response

The ECS sets an overall tone of safety in the body so the organism can learn new things, express curiosity, rest, relax, eat, and evolve. A healthy ECS decreases our acute response to stress, maintains a healthy response to perceived threats, and allows the extinguishing of fear.

In a healthy person, the amygdala, hippocampus, and prefrontal cortex are full of CB1 receptors and have high levels of endocannabinoids that regulate the tone of the nervous system. The amygdala helps create emotional memory and adapt to fear. It is regulated by the prefrontal cortex and the hippocampus, so keeping these areas healthy and full of synaptic connections is in our best interest. This regulation is diminished in conditions of the psyche. Chemically, the ECS carries out its functions with endocannabinoids binding to CB1, CB2, and TRPV1 receptors, regulating serotonin and glutamate levels.

The physiological mechanism of fight-or-flight is a complicated piece of equipment we are endowed with, and it serves our survival well. The ability to channel resources toward defending the body or running away is appropriate when we are under attack. The issue we face is not the hardwiring but that we can't log out or disengage and the neurochemistry keeps sending danger messages when in fact there is no danger.

The fight-or-flight cascade begins when the amygdala senses danger in conjunction with changes within the endocannabinoid system. AEA released within the brain promotes a sense of safety and well-being. In order for us to mount our fight-or-flight stress response, AEA levels must immediately drop (the chemical equivalent of "Hey, all is not well!"). AEA levels drop when the amygdala releases corticotropin-releasing hormone (CRH). CRH increases the amount of the enzyme FAAH, which degrades AEA. The stress response proceeds through the hypothalamic-pituitary-adrenal axis when we release norepinephrine (adrenaline) and cortisol and are able to fight off anything our amygdala has perceived as threatening. Eventually, 2AG levels go up as cortisol levels go up, usually 20 minutes to an hour after the threat has passed. The rise in 2AG shuts down the stress response, and our feelings of safety return. 2AG, acting at the forebrain (cerebrum, hypothalamus, and thalamus), also allows us to habituate to a repeated stress; eventually we are able to adapt to it without entering into a fight-or-flight mode.

Mounting evidence exists that early stress and trauma can change the ECS's stress response. Chronically depressed and anxious people have fewer CB1 receptors and lower levels of endocannabinoids, making the fight-or-flight response more difficult to modulate.

According to the Western medical model, people with conditions of the psyche (depression, anxiety, and PTSD) have one or more of the following: decreased modulatory role of the hippocampus and prefrontal cortex over the amygdala, a decreased production of serotonin or decreased binding of serotonin to its 5-HT1A receptor, or increased binding of glutamate to its receptor. Most pharmaceutical interventions target serotonin or glutamate levels. We will use this model of intervention when examining the role of cannabis in the conditions of the psyche. But first, let's take a look at each of these neurotransmitters separately.

SEROTONIN

One function of serotonin is to modulate the amygdala's response to stressful situations. There are high levels of serotonin receptors in the modulatory regions of the prefrontal cortex and hippocampus. Pharmaceutical antidepressants and antipsychotics work via the serotonin receptors.

Micromolar doses of CBD inhibit the degradation of tryptophan, the amino acid precursor for serotonin. Further studies have shown that CBD may be beneficial in diseases associated with immune activation and inflammation that lead to decreased levels of tryptophan. Evidence also suggests a mechanism whereby CBD's ability to decrease inflammation and neutralize free radicals and ROS by activating glial cells could prevent neuronal changes leading to depression.

CBD and CBDA (100 times more powerful CBD) bind to serotonin receptors. CBD also binds to dopamine receptors. CB1 receptor agonists AEA and THC activate an increase in serotonin. Studies have shown that the efficacy of AEA is equal to that of pharmaceuticals. Binding of CB1 receptors by its agonists increases the release of serotonin, dopamine, and norepinephrine (all linked to mood) from neurons and decreases their reuptake. CBG is a serotonin antagonist and should not be part of a formula for conditions of the psyche.

GLUTAMATE

Decreasing glutamate is a pharmaceutical target of antipsychotic drugs and obsessive-compulsive disorders. The binding of the endocannabinoids and THC to CB1 and TRPV1 causes a decrease in glutamate. CBD, by increasing AEA and binding to TRPV1 directly, decreases glutamate. CBD blocks THC-induced psychosis via this mechanism.

CBD and the Amygdala

CBD is a real superstar of amygdala modulation. It increases blood flow to and functioning of neurons in the prefrontal cortex. CBD also stimulates neurogenesis in the hippocampus and increases synaptic connections. All of this allows one to override an overactive amygdala in a fight-or-flight crusade to keep you alive.

Functional MRI shows that CBD activates the regions that modulate the amygdala (prefrontal cortex and hippocampus), quieting the amygdala. CBD and THC show enhanced reward-system activation, and both increase the production of neuron growth factors, which increases the size of the amygdala-modulating hippocampus.

CBD increases the blood-brain barrier function, and there's evidence of diminished barrier function in depressed individuals. Binding of THC and endocannabinoids to CB1 receptors increases neuron growth factors, increasing neurogenesis in the hippocampus. Further benefits of THC binding to CB1 receptors are inducing euphoria and focusing the mind on the present. Remember, when unopposed, THC can cause anxiety and, at higher doses, psychosis, so when working with the psyche and THC, lower doses are recommended in conjunction with higher-dose CBD.

Anxiety

Anxiety disorders are the most common group of mental illnesses in the United States, and relief from anxiety is one of the most commonly reported motives for working with cannabis. Anxiety is a crucial and necessary emotion. The tension and worry we feel when faced with an unknown or fearful situation causes us to act for our survival. Avoiding predation served our ancestors. It allowed them to survive and reproduce. But continuing to worry after an immediate threat has passed is not beneficial to us. It sets up neural pathways that keep a person in a constant state of fight or flight, never allowing for full rest or relaxation and setting the stage for disease.

Anxiety is a disease of the mind, the emotional body, and the spirit body. To bring about true transformation, we need to work with all components of the person; otherwise we merely manage symptoms. Cannabis has the unique ability to work at all three levels.

Working with cannabis for anxiety requires us to commit to our own learning and healing while we temporarily find relief from maladaptive neural ruts of anxiety. It is good and right to receive relief, to feel what life is like when not constantly listening to the narrative of anxiety. This relief is the gift cannabis offers us. Our co-creative role in this relationship is to commit to our own healing and learning and to create new neural grooves. Curiosity is a good ally.

Possible Misuse for Anxiety

One area for possible misuse of cannabis and its ability to decrease anxiety is when feelings of anxiety signal the need for a significant change in a person's life. Our emotions are chemical messages carrying vital information for our health and well-being. If you are in a stressful and demanding situation that requires a lifestyle change, utilizing cannabis (or any anxiolytic herb or drug) will only manage the symptoms, not the underlying causes. Paying attention to our symptoms is a way to tune in to the change we may need in our life — whether you're changing jobs, leaving an unhealthy relationship, or grieving over the death of a loved one or the current state of the world. Taking an herb or a pharmaceutical that dulls the message is not always helpful. Just because it's herbal doesn't mean it can't be misused to manage the symptom rather than address the cause.

Cannabis Withdrawal

Anxiety may manifest during cannabis withdrawal. During the first three days of abstinence, people may experience nervousness, restlessness, irritability, and sleep disturbances, the very things they may have been hoping cannabis would help with. These symptoms usually peak within one week. After 28 days, all withdrawal symptoms disappear.

Mechanism of Action

Common pharmaceuticals used to work with anxiety disorders are serotonin reuptake inhibitors, tricyclic antidepressants, and benzodiazepines (which work to increase the neurotransmitter GABA). In anxiety it is believed the amygdala is overactivated by a decreased amount of serotonin or GABA.

Research on Effectiveness

There are limited human clinical trials for cannabis and anxiety, yet relief from anxiety is one of the most common reported motives for people using cannabis. The National Academies of Sciences, Engineering, and Medicine released a report in 2017 indicating "limited evidence" for the effective use of cannabis for anxiety.

Two separate studies focused on the anxiety of public speaking tested the efficacy of isolated CBD and found that doses of either 300 mg or 600 mg decrease anxiety. An animal study found chronic rather than acute administration of CBD works better for reducing anxious behaviors. An observational study of veterans who self-administered cannabis found a 75 percent decrease in reliving traumatic experiences and avoidance.

Serotonin reuptake inhibitors, tricyclic antidepressants, and benzodiazepines are all strong pharmaceuticals with adverse effects.

Less than half of people using these drugs show complete and sustained remission of symptoms, even after long-term treatment.

Dosing

Given the nature of anxiety and the tendency for THC to cause anxiety, cultivar selection is crucial. High-CBD plants with very little THC are recommended to start. The terpene profile should also address specific anxiety symptoms. For example, if a person is anxious with low energy (depressed), stimulating terpenes like terpinolene and pinene are good choices, as are the uplifting floral (linalool) and citrus (limonene) terpenes. Higher myrcene levels will tend to be more sedating and grounding.

A good protocol is to start with 2.5 to 5.0 mg of high-CBD tincture every 6 to 8 hours. Increase the dosage to 50 mg if needed over a few weeks. When at an anxiolytic dose of CBD, THC can be added, starting with 1 mg of THC per dose. Remember the rule of minimum effective dose: more is not necessarily better. If you gain relief with 5 mg of CBD there's no reason to use more.

Depression

Depression is defined as a feeling of sadness and/or loss of interest in activities once enjoyed, low mood, hopelessness, trouble concentrating, and pain. It affects 6.7 percent of adults, with 15 percent of men and 30 percent of women experiencing depression at some time during their life. There are many theories about how and why people develop depression — from genetic predisposition to trauma to environmental factors. Other illnesses such as chronic fatigue and hypothyroidism, and some pharmaceuticals, can mimic or cause depression.

Cannabis use for melancholia dates back 150 years in America. Fifty percent of people using pharmaceuticals for depression do not achieve long-term relief. Given that statistic, it's not surprising that 30 to 50 percent of people using medical cannabis said they came to cannabis for relief from depression.

Mechanism of Action

The classic pathophysiology of depression is thought to stem from three areas: neurochemical deficits (serotonin levels), neurodegeneration of the hippocampus and prefrontal cortex, and disturbances in the hypothalamic-pituitary-adrenal axis (the pathway of the fight-or-flight response). Chronic inflammation, a disrupted gut biome, and dysregulation of the immune system can also be causative factors.

A heightened stress response initiated by the amygdala's perception of threat translates to an increased fight-or-flight response. This is often coupled with impaired functioning of the prefrontal cortex and hippocampus, which modulate the amygdala's response.

Neurotransmitters that signal reward and pleasure and allow us to concentrate, such as dopamine, serotonin, and norepinephrine, may be at lower levels in depressed people. Decreases in motivation, pleasure, and reward can stem from a decrease in neurological firing in the reward centers of the brain.

It's interesting to note that depressed, anxious, and stressed people have a smaller hippocampus than healthy individuals. Remember, the function of the hippocampus is to regulate mood, memory, and the amygdala.

The ECS and Depression

The endocannabinoid system relates to depression in a few ways. People with a genetic variance for fewer CB1 receptors may be predisposed to develop depression, especially after stressful events. Mice bred without CB1 receptors become sensitive to depression, have a heightened amygdala response, experience less reward, have decreased serotonin levels and neurotrophic factors (which grow nerves), and have smaller hippocampal regions.

To maintain psyche health, we need a healthy, functioning ECS. You'll remember that CB2 receptors, when bound to by 2AG and AEA, decrease inflammation, increase neuron growth factors, and modify the immune system. Tonic release of AEA within the brain promotes emotional homeostasis; 2AG shuts down the fight-or-flight process. With these functions in mind, it's interesting to note that chronically depressed people have less 2AG and AEA and fewer CB1 receptors able to respond to the decreased amounts of 2AG and

AEA. This results in a decreased ability to navigate the stresses of life. A smaller hippocampus and decreased firing from the prefrontal cortex results in a diminished ability to calm the overactive amygdala.

Mounting evidence exists that early stress and trauma can change the ECS's response to stress. Functional MRI shows that CBD activates the prefrontal cortex and hippocampus, quieting amygdala activation. CBD and THC show enhanced activity in the reward system region of the brain, and both stimulate the production of neuron growth factors, increasing the size of the amygdala-modulating hippocampus.

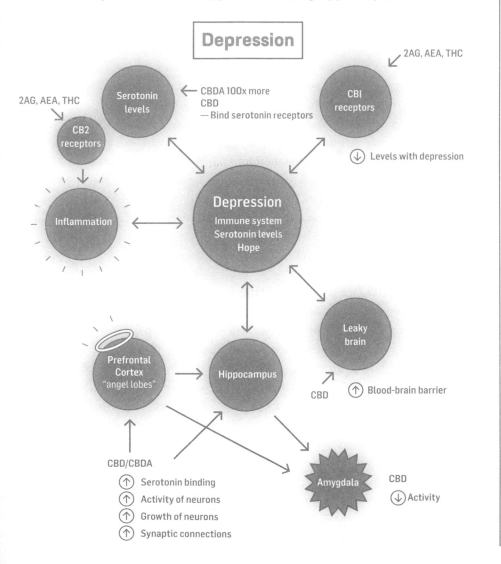

Cannabis and Depression

Cannabis interacts with depression through a multipronged approach. It activates modulatory regions for the amygdala, calms the excited amygdala, increases mood-controlling neurotransmitter levels, binds to serotonin and dopamine receptors, and stimulates neurogenesis. The Western medical model works with depression mostly by one mechanism: utilizing reuptake inhibitors, which keep serotonin or norepinephrine in the synapse longer.

CBD increases the blood-brain barrier function, and there's evidence of diminished barrier function in depressed individuals. THC binding to CB1 receptors increases the release of serotonin, dopamine, and norepinephrine from neurons and decreases their reuptake. It also increases neuron growth factors, which increases neurogenesis in the hippocampus. CBD and CBDA (100 times more potent) both bind to serotonin receptors. CBD also binds to dopamine receptors. Linalool, found in cannabis, is another uplifting terpene, both anxiolytic and antidepressant.

Some of what we know about the ECS in relation to depression and anxiety comes from pharmaceuticals gone wrong. The drug rimonabant, approved by the FDA and put on the market for weight loss, was pulled when people using this CB1 receptor blocker were 2.5 times more likely to become depressed and two times more likely to commit suicide.

Research on Effectiveness

Rodent models of depression showed THC and CBD equal in efficacy for treating depression to pharmaceutical antidepressants. One study of depressed people showed 200 mg of CBD taken for 10 weeks showed an increase in hippocampus size, while no change was seen in healthy people. Limonene increases dopamine and serotonin levels in the prefrontal cortex; in a Japanese study, 9 out of 12 depressed people found enough relief with limonene to stop using pharmaceutical antidepressants.

Dosage

Start with 5 mg of a high-CBD strain every 6 hours. Remember, some people find CBD stimulating; they should not take a dose after

5:00 P.M. A high-THC strain in doses of 2.5 to 5.0 mg lifts mood. High doses of THC can be anxiety promoting, so it's best to stay at this low dose range.

Cultivars high in pinene should be avoided because some evidence exists that pinene may interfere with memory remodeling and fear extinction. Cultivars high in myrcene should also be avoided because they tend to sedate.

Posttraumatic Stress Disorder (PTSD)

PTSD is a disorder some people develop after an intense experience of terror or helplessness. Symptoms of PTSD can include insomnia or nightmares, anxiety, an inability to tolerate frustration, anger issues, flashbacks, and hyperarousal.

Everybody experiences stressful or dangerous events at some point in life, but with time we usually forget to be afraid of them. Normally we can eventually learn to dissociate components of the event (say, a sound or smell) with the painful event itself. Healthy nervous systems don't go on high alert with the slightest memory of the event. The ability to forget to be afraid and the uncoupling of the event with the strong emotions and sensations that occurred during it are all functions of a healthy endocannabinoid system and a healthy, functioning relationship between the amygdala and its mod-ulators, the prefrontal cortex and the hippocampus.

People with PTSD are unable to forget to be afraid. They experience memories of a traumatic event as if the event is happening in present time. Benign stimuli can trigger painful memories of the event. Evidence exists that endocannabinoid deficiency and an atrophied hippocampus contribute toward developing PTSD after a traumatic event.

Cannabis and PTSD

The primary goals in the treatment of PTSD are the extinction of the fear memory and the uncoupling of the fear memory from triggers. Another goal is a sense of safety and well-being. Cannabis is well equipped to support both these goals. However, even with all the benefits of cannabis, it is important to incorporate other modalities of healing along with cannabis when addressing PTSD.

Mechanism of Action

PTSD sufferers show changes in the structure and function of the amygdala with a concurrent decrease in hippocampus and prefrontal cortex function. They also show a decrease in endocannabinoids and diminished CB1 receptor numbers in those areas.

Research on Effectiveness

Cannabis has been shown to be helpful with PTSD in a number of ways. CBD has been shown to:

- Decrease defensive behaviors in mouse models of PTSD

- Increase neurogenesis in the hippocampus

- Increase ECS function via FAAH inhibition, which results in more binding of AEA to CB1 receptors

- Decrease learned fear expression

- Increase fear extinction

- Decrease stress-related anxiety and anxiety after stressful events

THC helps with:

- Sleep and the reduction of nightmares

- Increasing neurogenesis and improving mood at low doses via binding to CB1 receptors

- Decreasing nightmares; taking 0.5 to 3.0 mg of THC before bed showed a 72 percent decrease in nightmares in PTSD sufferers

Dosage

Take 5 to 10 mg CBD every 6 hours to start. If more relief is needed, use a 10:1 ratio of CBD:THC, 5 to 10 mg CBD plus 0.5 to 1.0 mg THC every 6 hours; 5 mg or less of THC smoked twice per day helps to increase sleep and decrease nightmares. Strains high in pinene should be avoided, as they may interfere with memory remodeling and fear extinction.

Schizophrenia

Schizophrenia is a chronic neurodevelopmental disorder with three main symptom domains: positive symptoms, negative symptoms, and cognitive defects. Positive symptoms are psychotic behaviors not generally seen in healthy people, including hallucinations, delusions, and paranoia. Negative symptoms are disruptions to normal emotions and behaviors: social withdrawal, flattened emotional expression, and lack of motivation. Cognitive defects can include an impaired ability to understand information, problems with working memory, and trouble focusing. Existing pharmaceutical medications offer minimal relief from positive or negative symptoms and no benefits (or negative effects) to cognitive deficits. Adverse effects of pharmaceuticals include diabetes mellitus and weight gain.

Mechanism of Action

Schizophrenics are more sensitive to THC's psychomimetic effects — those that mimic the disease. Psychomotor skills, short-term memory, working memory, verbal learning, and decision making are all negatively affected. The regions in schizophrenics' brains responsible for information processing show lower levels of GABA and glutamate. When THC binds to CB1 receptors, glutamate and GABA levels decrease even more, which may be the basis for THC sensitivity. Pharmaceutical-induced GABA deficits replicate the psychomimetic effects of THC.

Cannabis rich in CBD, on the other hand, has been shown to alleviate symptoms with its ability to increase AEA by decreasing the degrading enzyme FAAH. CBD has also been proven to be anxiolytic and antipsychotic via different mechanisms.

Research on Effectiveness

Rodent models of schizophrenia showed improvement of both cognitive impairment and social interaction deficits with a 680 mg dose of CBD two times per day. One human study showed that 200 mg CBD four times per day improved both positive and negative symptoms as effectively as pharmaceuticals, without the adverse effects. Another human study showed an increase in positive symptoms and cognitive performance when 1,000 mg CBD per day was co-administered with existing medications.

Dosage

Based on the literature, dose 200 mg CBD four times per day.

Insomnia and Other Sleep Problems

Sleep is essential for our health and well-being. Lack of sleep is linked to chronic disease, including chronic inflammation and pain. Our bodies, brains, and psyches need 8 to 9 hours of sleep per day to rest, rebuild, recover, assimilate the day's events, and clear the circuitry of our neurochemistry so we can start fresh. Just one night of poor sleep decreases reaction time, decreases cognitive performance, lowers energy, aggravates pain and inflammation, and stimulates overeating.

Insomnia, the inability to fall asleep or stay asleep, ranks as a top reason people turn to cannabis. It's not surprising that people turn to cannabis for help with sleep, given that 50 percent of insomnia sufferers are dissatisfied with pharmaceutical interventions. Moreover, many sleep drugs have serious long-term adverse effects, including dependency. Pharmaceutical sleep aids are meant to be used short-term, and when people stop using them, they can experience rebound insomnia and anxiety. And many pharmaceuticals used for other conditions can cause insomnia or sleep disturbances, especially antidepressants. Because inflammation and endocannabinoid deficiency are linked to many chronic diseases, and sleep helps in both of these situations, getting consistent, sound sleep is not a luxury but an essential component to our health.

Cannabis and Sleep

Cannabis helps with both anxiety and depression, and people feeling less anxious or depressed are better able to sleep. However, here are some things to keep in mind:

◆ Short-term cannabis use helps with falling asleep, but the effect weakens over time due to tolerance.

◆ Excessive THC may decrease restful normal sleep and leave people feeling not well rested in the morning.

◆ Chronic users of cannabis spend less time in the deep restorative phase of sleeping.

◆ Withdrawal from chronic use of cannabis can cause sleep disturbance and anxiety, the very symptoms someone might initially begin to work with cannabis to alleviate. Withdrawal symptoms disappear within one to three weeks.

Mechanism of Action

The sleep cycle — falling asleep, staying asleep, waking up, and remaining awake — is governed by complex circadian rhythms and our endocannabinoid system. The ECS functions to maintain wakefulness *and* to generate sleep. (Doesn't that sound delicious!) AEA levels increase at night, quieting the nervous system so we can fall asleep. 2AG levels increase during the day to maintain wakefulness. AEA and 2AG both bind to CB1 receptors to enhance wakefulness during the day and maintain deep (non-REM) delta sleep at night (a highly restorative sleep phase).

Research

THC can be initially stimulating, but after 90 minutes tends to be sedating. All studies on sleep and cannabis show THC to be most effective either by itself or in a 1:1 ratio with CBD. CBD at low doses is stimulating and helpful for maintaining wakefulness. High-dose CBD (160 mg in one study) can help increase length of time sleeping and decrease sleep interruptions. One study also showed high-dose CBD to decrease REM behavior disorders (the acting out of dreams).

CBD also decreases anxiety and helps excessively fatigued people fall asleep without providing too much stimulation for well-rested people.

CBN, the oxidized product of CBD and THC, is sedating and potentiates the sedative action of THC. Sleep-promoting terpenes include terpinolene, nerolidol, phytol, linalool, and doses of myrcene above 0.5 percent, because levels below that are mildly stimulating.

Forty percent of insomniacs are also depressed or anxious. Ninety-three percent of anxious or depressed people sleep better with cannabis. Many studies on cannabis and pain show an added benefit of helping with sleep. Pain levels decrease with increased sleep due to decreased substance P levels.

Dosage

Dose and delivery method are important in helping with sleep. In general, 1:1 extracts or high-THC extracts work better for sleep in the short term. For most people, CBD after 5:00 P.M. is too stimulating and should be avoided unless in a 1:1 ratio with THC. The dosing range is 1.5 to 15 mg of THC. Here are some specific dose recommendations.

FOR TROUBLE FALLING ASLEEP

* Tincture 20 minutes before bed

* Inhalation method 1 hour before bed

FOR WAKING IN THE NIGHT

* Oral/edible 1 hour before bed

A note about edibles here. Remember, go low and slow with edibles, especially if you are new to cannabis. THC's psychoactive and intoxicating properties may interfere with restful sleep.

Nausea and Vomiting

Nausea is a feeling of sickness with an inclination to vomit. Vomiting is a protective mechanism that is easily provoked to ensure that you

SELECTING CULTIVARS FOR SLEEP (LEARN FROM MY MISTAKE!)

Cultivar selection for sleep is crucial because the cannabinoid ratio and terpene profile could keep a person highly alert and up all night doing math calculations! I experimented with a high-CBD strain to see how well it promoted sleep. My *"But it's only CBD!"* attitude was misguided. My partner and I tried a very low dose in the evening to help with our various aches and pains and to help us sleep. Instead we were up all night!

expel toxins quickly; when you eat something poisonous, getting it out of your body quickly offers the most protection. Delayed-onset nausea and vomiting, on the other hand, usually appears hours to days after the introduction of something noxious. This most often occurs during chemotherapy.

Delayed-onset nausea and vomiting are difficult to treat with conventional pharmaceuticals. This complex process can be initiated in the brain, the gut lining, or the inner ear, and it's caused by a wide range of conditions: external stimuli like poison, toxins, or odors; and many pharmaceuticals, including antibiotics and opiates.

Cannabis, Nausea, and Vomiting

Cannabis helps with nausea and vomiting via three mechanisms: (1) CBD boosts AEA; low levels of AEA lead to nausea and vomiting; (2) THC and CBN bind to CB1 receptors, which decreases nausea and vomiting; and (3) CBDA and CBD bind fully to 5-HT1A receptors, and THC is a partial agonist.

Low-dose CBD and CBDA binding to serotonin receptors mediates nausea and vomiting and also helps mediate anxiety, especially in people with anticipatory nausea and vomiting.

Mechanism of Action

The endocannabinoid system regulates nausea and vomiting. A high density of CB1 receptors are found in the vomit centers of the brain, the gut lining, and the inner ear. Lower levels of AEA and 2AG or low levels of CB1 receptors lead to nausea and vomiting. This may also play a part in motion sickness.

Serotonin is a neurotransmitter involved in the mechanics of nausea and vomiting, specifically when binding to its 5-HT1A receptors. Binding at this receptor is also anxiolytic, antidepressant, and provides help with migraines.

Research on Effectiveness

Cannabis has been proven repeatedly to alleviate nausea and vomiting, even in hard-to-treat anticipatory cases. In the broad study by the National Academies of Sciences, Engineering, and Medicine, cannabis, when used as an antiemetic in the treatment of chemotherapy-induced nausea and vomiting, received the highest ranking of "conclusive evidence for therapeutic use."

Whole-plant medicine ranks as most effective. THC working at the CB1 receptors combined with CBD or CBDA binding to the serotonin 5-HT1A receptors both decrease nausea and vomiting. CBDA has been shown to be *100 times* more effective than CBD and doesn't show the potentiation of nausea and vomiting that CBD does at higher doses. Because of CBD's potential to increase nausea and vomiting at higher doses in animal models, I recommend keeping the dose low (below 2,000 mg, which is quite high).

It is worth noting we are in the early stages of understanding CBDA, but given that it is 100 times more powerful at binding to serotonin receptors, it might be excellent for relieving nausea and vomiting and worth adding into formulations. THCA has also shown antiemetic and antinausea properties.

Dosing

Dosing for THC has been studied more than CBD or CBDA. For acute nausea, start at a 2.5 mg dose of 1:1 THC:CBD — or replace some CBD with CBDA — and if needed, work up to the sweet spot of 10 to 12.5 mg for symptomatic relief. This can either be inhaled or used as an oral tincture.

For chemo-induced nausea, start two weeks before chemotherapy and work up to the sweet spot of 10 to 12.5 mg per day. The cannabis-naïve should start at 2.5 mg per day and work up to 10 mg per day to avoid intoxication. Avoid cultivars with CBG because it is a 5-HT1A receptor antagonist.

Conditions with an Endocannabinoid-Deficiency Component

Although irritable bowel syndrome (IBS) and migraine headaches may not appear to have anything in common, they do share one common causative factor: deficiency in the ECS. So while we may have different herbal protocols for helping with these conditions, all clients will benefit from boosting the ECS.

Irritable Bowel Syndrome (IBS)

IBS, also sometimes called spastic colon, is characterized by gastro-intestinal (GI) pain, spasms, discomfort, and diarrhea or constipation. It can be brought on by anxiety, specific foods, overeating, and an unhealthy gut biome. IBS is the most frequently diagnosed condition by gastroenterologists in the United States, at 10 to 15 percent of GI patients.

The endocannabinoid system maintains tonic control of gastro-intestinal motility and modulates secretion, inflammation, and pain along the GI tract. Thus, maintaining healthy levels of endocannabinoids or modulating this system using cannabis makes physiological sense for working with IBS.

Mechanism of Action

When there is an injury or chronic inflammation of intestinal tissue, the body increases production of inflammatory cytokines. Cytokines signal an increase in smooth muscle contraction along the digestive tract, which increases fecal transit time and decreases a person's ability to digest and absorb food. In response, the endocannabinoid system increases production of AEA, CB1 receptors, and CB2 receptors to alleviate the inflammation.

An individual with endocannabinoid deficiency syndrome (ECDS) loses some ability to upregulate the ECS to compensate for the inflammatory injury to the GI tract. So a partial mechanism of action for alleviating IBS is boosting the endocannabinoid system.

CB1 receptors are found within the enteric nervous system and in nonneuronal tissue, including intestinal mucosal cells, enterocytes (cells lining the intestine), immune cells, and enteroendocrine cells. Binding of the CB1 receptors by AEA, THC, or other agonists delays gastric emptying, decreases peptic acid secretion, and slows peristalsis.

CB2 receptors are found on immune cells along the GI tract and neurons of the lining of the tract. Binding of CB2 agonists to CB2 receptors modulates nausea, pain, and inflammation induced by increased transit time. TRPV1 nerve fibers, which are found at 3.5 times normal levels in IBS sufferers, are responsible for visceral hypersensitivity and pain. High-CBD strains boost endogenous AEA and directly desensitize TRPV1 nerves, helping to alleviate the discomfort.

Research on Effectiveness

Limitations of the few clinical trials with humans include (1) the use of synthetic cannabinoids in isolation; (2) single-dose administration rather than a longer regimen; and (3) the very few numbers of clinical trials.

Cannabis was the first effective intervention for the intense secretory diarrhea associated with cholera in the 19th century. Mouse models of cholera have proven that AEA binding to CB1 receptors alleviates secretory diarrhea. Pharmaceutical interventions targeting the 5-HT3 and 5-HT4 receptors were discontinued due to the adverse effects on the cardiovascular system and the development of ischemic colitis in people.

Conventional treatment utilizing anticholinergics, opioids, and antidepressants do not offer much relief. In 34 out of 42 studies on IBS, beneficial effects of probiotic supplementation were shown for one or more of the following: pain, discomfort, bloating, and distension. A healthy population of lactobacillus acidophilus stimulates an increase in mRNA for CB2 receptors in the gut.

Dosage

I recommend dosing with 1:1 cultivars higher in anxiolytic and antispasmodic terpenes. If alcohol exacerbates symptoms, use infused olive or MCT oil rather than alcohol tincture.

Migraine

Migraine presents as a hemicranial, beating headache that can be recurring and usually lasts from 4 hours to 3 days. It is often accompanied by nausea, vomiting, and sensitivity to light (photophobia) or sound (phonophobia), and it is sometimes preceded by an aura and is often followed by fatigue. Fourteen percent of Americans suffer migraines, with women having a threefold chance over men of experiencing them. Migraine may have several possible etiologies, including endocannabinoid deficiency, decreased serotonin levels, and trigeminal vascular system dysfunction.

Mechanism of Action

Cannabis works with migraine via four known mechanisms: (1) increasing serotonin levels or acting as an agonist at serotonin receptors; (2) quieting overactivity of the trigeminal vascular system; (3) acting directly as an analgesic; or (4) boosting a deficient endocannabinoid system.

SEROTONIN

Low serotonin levels (40 percent below normal) are found in migraine sufferers. CB1 receptor agonists AEA and THC directly activate an increase in serotonin. Accepted pharmaceutical migraine interventions at serotonin receptors 5-HT1A (for potentiation) and

5-HT2A (for inhibition) can be reproduced with AEA. AEA also antagonizes the 5-HT3 receptor, known for emesis and pain. CBD and CBDA directly activate 5-HT1A receptors as full agonists, while THC acts as a partial agonist initiating the effects of serotonin. CBG is an antagonist and should not be part of a formula for migraines.

TRIGEMINAL VASCULAR OVEREXCITATION

The trigeminal vascular overexcitation theory of migraine begins with a stimulating event (such as stress, trigger foods, or hormone levels) that activates axons of the trigeminal nerves. The activated nerves release vasoactive neuropeptides such as SP, neurokinin A, and calcitonin gene-related peptide, which cause neurogenic inflammation, vasodilation, leakage of plasma proteins, and mast cell degranulation (more inflammatory cytokines released and leaking into the brain). This vasodilation and the leakage of cytokines is thought to be responsible for the pain of migraines.

CBD strengthens the blood-brain barrier and lowers inflammatory cytokines. CB1 agonists AEA, 2AG, and THC inhibit A and C fibers of the trigeminal nerves, which decreases overall neuronal activity or the cortical spreading depression found in migraine sufferers.

ANALGESIA

AEA is tonically active, and its continued release in the periaqueductal gray (PAG) area (pain region) of the midbrain provides analgesia and prevents dilation of blood vessels of the dura in response to the release of vasoactive neuropeptides. It also prevents the hyperalgesia of phonophobia and photophobia that sometimes accompanies migraine. Deficiency of AEA prevents analgesia and vasodilation (the ability to vasodilate and vasoconstrict are necessary for healthy blood flow), and indeed in numerous studies migraine sufferers showed lower AEA levels due to increased functioning of the degradation enzyme FAAH and the AEA transporter. CBD decreases transporter function and FAAH activity to increase AEA levels.

TRPV1 receptor activation has also been implicated in the hyperalgesia of migraine. Desensitizing it with either AEA or CBD would be an effective intervention. Intranasal capsaicin has been used for

migraine by brave souls. Oral CBD would be a less noxious and friend-
lier intervention.

ENDOCANNABINOID DEFICIENCY

Because the endocannabinoid system maintains tonic control of anal-
gesia/hyperalgesia at the PAG area of the brain, low levels of AEA predict a tendency for migraines because this area is thought to generate
migraine and pain in the brain. Photophobia and phonophobia are
expressions of sensory hyperalgesia, which is modulated by the PAG.
Studies where CB1 receptors are blocked show increased hyperalgesia.
AEA is tonically released to modulate the trigeminovascular system,
causing a decrease in vasodilation induced by calcitonin gene-related
peptide and nitric oxide (NO), two vasoactive neuropeptides released
upon nerve stimulation and known to induce neurogenic inflammation
that leads to migraine. AEA causes a dose-dependent vasodilation as
well. Migraine sufferers tend to have lower serum AEA levels due in
part to increased FAAH enzyme levels and AEA transporter function.
Cerebrospinal levels of AEA are also lower in chronic migraine sufferers. THC at low doses stimulates AEA biosynthesis.

Research on Effectiveness

The main treatment for migraines in Europe and the United States
from 1843 to 1943 was cannabis. A Colorado-based survey of
120 self-reporting cannabis users for migraine found a decrease from
10.4 headaches per month to 4.6 per month. Numerous studies find
evidence of endocannabinoid deficiency in migraine sufferers.

Dosage

Lifestyle change is a large factor in managing migraine. Removing
triggers and boosting the ECS form the foundation of working with
migraine. Consider cultivars devoid of CBG and containing linalool and
other soothing terpenes. Also consider CBDA for its increased capacity
to bind to serotonin receptors. Mix CBD and CBDA in a 1:1 ratio. Work
up to 80 mg total per day. Then work in a low-dose THC tincture. Use
drops of 10mg/mL THC tincture for accute migraine relief.

Conditions of the Immune System

Cancer and multiple sclerosis (MS) are grouped together here because at the root of both diseases is immune system dysfunction. In the case of cancer, the immune system cannot recognize cancer cells as foreign and kill them or it can't keep up with killing them. With MS, the immune system is attacking its own cells, causing disease.

Cancer

Cancer is a disease in which abnormal cells divide out of control. These abnormal cells no longer carry out their intended function but rather become freeloader cells, consuming nutrients, growing, and taking up space. In some cases, the cancer can metastasize, spreading to other tissues and consuming more resources. The issue for the healthy cells is that the cancer cells (I like to think of them as blob cells) don't do the job of the tissue, and they consume resources the healthy functioning cells need. Further, cancer cells secrete chemicals that promote blood vessel growth (angiogenesis) so they can continue consuming and growing. Eventually, the tumor or tumors become so big and wide-spread that there aren't enough resources for healthy tissue to carry out their functions. When we die from cancer, we die from the inability to perform the necessary tasks of the organ or organs with the cancer.

Our bodies make 10,000 DNA "mistakes" every day — we are essentially always precancerous. Luckily, we have ways of keeping these rogue cells in check: the immune system and the endocannabinoid system. Our immune system destroys rogue cells, and the ECS decreases tumor cell growth.

Mechanism of Action

Our natural cancer prevention has a four-tier response. Our body can (1) kill rogue cells directly; (2) induce autophagy (cell eating) and apoptosis (natural cell death) of cancer cells; (3) prevent angiogenesis, metastasis, and tumor invasion; and (4) slow tumor cell growth.

The endocannabinoid system is also antitumorigenic. Endocannabinoid levels increase in tumors as a homeostatic mechanism to decrease their growth. Endocannabinoid degrading enzymes are downregulated (prolonging the effects of endocannabinoids), indicating the body's natural healing response to cancer. Tumor growth rates decrease when fewer degradation enzymes of ECS chemicals are present.

NATURAL KILLER CELLS

The body's immune system in its wisdom has developed a method of removing rogue cells. It produces natural killer cells (NKCs) that circulate in the blood, lymph, and connective tissue looking for foreign cells, viral particles, harmful molecules, or your own cells gone rogue. When NKCs find cancer cells, they send out cytokine signals to other immune cells for help, then go about poking holes in the rogue cells to kill them.

CBD increases intercellular adhesion molecule (ICAM) expression in cancer cells. ICAM increases NKCs' adherence to cancer cells, thus increasing their effectiveness. CBD does not increase ICAM expression in healthy cells, unlike chemotherapy, which kills *all* rapidly dividing cells, healthy and cancerous alike.

AUTOPHAGY AND APOPTOSIS

Immune cells can also induce apoptosis (natural, programmed cell death) or autophagy in cancer cells. In the ECS, binding of CB receptor agonists on cancer cells induces apoptosis and autophagy via three different intracellular pathways: MAPK, PI3K, and inhibition of adenylate cyclase. Remember, endocannabinoids and THC bind to CB receptors. CBD, by inhibiting the degrading enzyme FAAH, increases AEA, which allows increased binding of CB receptors. CBD binding to GRP55 and TRPV1 receptors induces autophagy and apoptosis in tumor cells as well.

ANGIOGENESIS AND METASTASIS PREVENTION

THC also decreases angiogenesis, the process whereby cancer cells encourage blood vessels to grow toward them. If the blood supply is cut off, the cancer cells will wither and die and won't outcompete the healthy cells for resources.

THC inhibits angiogenesis by decreasing vascular endothelial growth factor and its receptors. THC also decreases the metastatic movement of cancer cells. In animal models, CBD also decreases metastasis. And CBD increases ICAM expression in unhealthy cells, which leads to decreased invasion by cancer cells.

SLOWING TUMOR CELL GROWTH

When endocannabinoids bind to CB receptors, cell growth decreases in many tumor cell lines, including glioma, melanoma, breast carcinoma, hepatic and pancreatic cancers, thyroid epithelioma, uterine carcinoma, biliary tract cancer, cervical carcinoma, colorectal carcinoma, gastric adenocarcinoma, leukemia, lung carcinoma, lymphoma, oral cancer, pancreatic adenocarcinoma, prostate carcinoma, and skin carcinoma. It's important to note that these are cell lines, cancers grown in petri dishes, not living tumors in humans. We can extrapolate that they would work in humans, but the research has not been done.

THC decreases cell growth and tumor growth by blocking epidermal growth factor (EGF). Cancer cells possess an unusually high number of receptors for EGF. Decreasing the amount of this growth factor makes it less available to bind to cancer cells, thus decreasing signals to keep growing.

CBD does not directly bind to CB receptors but acts via different mechanisms to decrease tumor growth. CBD has also been shown to decrease cancer cell viability in neuroblastoma, glioblastoma, melanoma, leukemia, colorectal, breast, lung, and prostate cancers.

POTENTIATING CHEMOTHERAPY MEDICATIONS

THC does all of this anticancer work while protecting healthy, noncancerous cells. It has also been shown, along with CBD, to potentiate chemotherapy meds. Potentiating chemotherapy medications means you can take less of them, which results in fewer adverse effects of the chemotherapy medications themselves.

The ECS-Immune System Relationship

After injury or disease, the body's response for healing is local inflammation. This can be shut down by increasing endocannabinoid production and/or increasing CB2 receptors. Both help decrease production of inflammatory cytokines and increase production of anti-inflammatory cytokines, ultimately decreasing inflammation. The level of CB2 receptor numbers on immune cells correlates with their activation state. When the body cannot stop inflammation, the inflammation becomes chronic. The ECS signals cytokine production, which directly signals activating or quieting of the immune system.

When endocannabinoids or phytocannabinoids bind to CB receptors, inflammation and the production of T cells (which make inflammatory cytokines) are both suppressed, T cell apoptosis is induced, and fewer immune cells migrate to and adhere to the area of injury. But if the inflammatory process shuts down too soon — for example, when there is a live pathogen or with certain cancers — increased damage may occur because the protective role of the immune system has been compromised. Understanding which cancer cells cannabis is effective against is important when working with cannabis to fight cancer.

There are also some cases where the ECS's suppression of the immune response can lead to an increase in tumor growth due to an increase of cytokines IL-10 and TGF beta and a decrease of IFN gamma. Here, the balance between the antitumor actions of CB agonists and their immunosuppressive actions may shift in the direction of an increase in tumor growth. This is especially true in tumors with low CB receptor expression. They are less sensitive to the anticancer actions of the CB receptor agonists on the tumor while the immune system is still susceptible to the suppressive effects of CB agonists, with a net effect of encouraging tumor growth. In short, not all cancers are stopped by cannabis. The antitumorigenic properties of CBD via noncannabinoid receptor mechanisms or other palliative benefits listed for cannabis could still occur.

The following is a list of tumors known to respond to cannabis: hepatocellular carcinoma, endometrial carcinoma, glioblastoma multiforme, meningioma, pituitary adenoma, Hodgkin lymphoma, chemically induced hepatocarcinoma, mantle cell lymphoma, neuroblastoma, melanoma, leukemia, breast carcinoma, thyroid epithelioma, uterine carcinoma, biliary tract cancer, cervical carcinoma, colorectal carcinoma, gastric adenocarcinoma, lung carcinoma, oral cancer, pancreatic adenocarcinoma, prostate carcinoma, and skin carcinoma.

Research on Effectiveness

Currently, cannabis is recognized by Western science and medicine for use in palliative (quality of life) cancer care to decrease nausea and vomiting, alleviate pain, stimulate appetite, elevate mood, and relieve insomnia.

Research in tumor cell line studies, in vitro animal studies, and human trials suggests cannabis can directly and indirectly decrease tumor growth. Cannabinoids in nude mice curb tumor growth in lung carcinoma, breast carcinoma, skin carcinoma, melanoma, thyroid epithelioma, lymphoma, and glioma, while normal glial cells were unaffected.

A 2018 review by the National Cancer Institute's Cancer Integrative, Alternative, and Complementary Therapies Editorial Board stated that cannabis contributes to pain relief for people living with cancer along with being an antiemetic, appetite stimulant, and sleep aid. All but 1 of 34 in vitro studies showed cannabis kills tumors selectively via antiproliferation, decreasing cell viability, and initiating cell death via toxicity, apoptosis, necrosis, and autophagy; cannabis is also anti-angiogenic and antimetastatic.

In a study of 21 people with recurrent glioblastoma multiforme taking Sativex (a 1:1 CBD:THC whole-plant extract) with temozolomide, the one-year survival rate went from 53 percent with a placebo to 83 percent using Sativex; the median days of survival was 550 with Sativex versus 369 with placebo.

In an Israeli study of 2,000 people with cancer, improvements in reported symptoms were:

- Nausea and vomiting: 91 percent

- Sleep: 87.5 percent

- Restlessness: 87.5 percent

- Anxiety and depression: 84.2 percent

- Pruritus (itchy skin): 82.1 percent

- Headache: 81.4 percent

About half of these patients reported a 35.1 percent decrease in concomitant medications. Of 344 people using opiates, 36 percent stopped and 46 percent decreased their daily dosages.

A study on the murine model of glioma in mice found a dramatic decrease in tumor growth in vivo with a 1:1 extract used concurrently with ionizing radiation. Further, 50 percent less THC and CBD were needed when they were combined in the extract. The factored equivalent dose in humans was 136 mg of each given three times over one week.

Studies in murine models of cancer with silenced CB1 receptors show accelerated cancer growth. After the same CB receptors are activated, a decrease in growth is seen in mouse colon cancer models.

Dosage

If working with palliative care, the dose is still MED of 1:1 full-plant extract. For specific symptoms (nausea and vomiting, alleviating pain, stimulating appetite, mood elevation, and relief from insomnia), refer to the dosing considerations listed for each separate condition in chapter 4.

Working with cancer-fighting aspects of cannabis is controversial, with limited human trials but many anecdotal reports. The dosing in animal studies ranges from 700 to 3,400 mg of THC per day for 10 days to 2 years. The regimen needs to be coordinated with a person's primary care physician. The protocol consists of a regimen working up to 1,000 mg per day of THC and 400 mg per day of CBD. This is one of the few situations necessitating resin extracts. Please do not attempt to implement this protocol by yourself. Supervision with a provider who is experienced working with this medicine is important; depending on your cannabis tolerance level, you will need supervision at home for at least a month. A 1,000 mg daily dose is extremely high, and most people must work up to it. Tolerance of a level to return to "unsupervised life" takes three to five weeks.

Immune System T Cells, Cancer, and Cannabis

The immune system, T cells in particular, is involved in the control, growth, and development of many cancer types. Endocannabinoids and the cytokine network communicate directly, "talking" to each other constantly. All immune cells express cannabinoid receptors and can up- or downregulate the number of these receptors depending on their level of activation. Understanding the immune functions of T cell lymphocytes will help map out the mechanism of action of endocannabinoids and cannabinoids.

There are five T cell subtypes: helper T cells, cytotoxic T cells, memory T cells, regulatory T cells, and natural killer cells.

Helper Ts (Th) Two different Th cells exist, Th1 and Th2. Th1 cells contain CD4 proteins on their surface and are sometimes referred to as CD4 cells. They are considered pro-inflammatory; they stimulate the healing process of inflammation and are crucial for effective immune responses against cancer. Cytokines that promote Th1 are IL-2, IL-12, and interferon gamma. IL-10 suppresses Th1. IL-2 also triggers T, B, and natural killer cell proliferation and activation. Th2 cells are considered anti-inflammatory. Cytokines that promote their growth and expression are IL-4, IL-5, IL-10, and transforming growth factor beta.

Cytotoxic Ts (Tc) Tc cells contain a specific CD8 protein on their surface and are often referred to as CD8 cells. Tc cells kill with specificity, but they need co-activation by Th cells to begin their work. Binding of CB2 by either THC or endocannabinoids suppresses cytotoxic T cell proliferation.

Memory Ts (Tm) Tm cells are also called "antigen-experienced cells" because they have been introduced to the specific foreign cell before and retain the ability to immediately react. Memory T cells allow for years of quick acting by the immune system to something it has previously been exposed to.

Regulatory Ts (Tr) Tr cells inhibit or shut down immune responses initiated by Th cells and are important in immune system health. Binding of CB2 by either THC or endocannabinoids upregulates regulatory Ts.

Natural Killer Cells (NKCs) NKCs klll any cell without a "me flag" (major histocompatibility complex marker) — a flag that most of your cells have unless they have become cancerous. NKCs are nonspecific and don't need to be co-stimulated by Th cells. NKC surveillance is crucial to preventing cancer.

Multiple Sclerosis

Multiple sclerosis is an autoimmune disorder involving specific reactive T cells, macrophages, microglia, and astrocytes within the central nervous system. The body's immune system begins to recognize the insulating fat surrounding the neuron (myelin sheath) as foreign and initiates inflammation. The secondary response occurs when specific T cells begin attacking the myelin, the cells that make myelin (oligodendrocytes), and the neurons themselves. This process leads to the demyelination of the neurons, making them less and less effective at conducting nerve impulses. The body can resynthesize myelin initially, but over time the neuron becomes scarred, giving the disease its name ("sclerosis" means "scarring"). Scar tissue, by definition, does not function like healthy tissue.

Because the disease targets neurons within different regions of the brain and spinal cord, there can be a spectrum of signs and symptoms depending on where the demyelination occurs. These signs and symptoms may include muscle spasms, tremors, ataxia (loss of muscle control), weakness or paralysis, constipation, and loss of bladder control.

Mechanism of Action

Under healthy conditions, the blood-brain barrier blocks immune system access to the cells within the central nervous system. In pathological conditions, the blood-brain barrier becomes leaky, and helper T cells enter the extracellular space and gain access to neurons. The T cells mistake the myelin as foreign material and begin making the inflammatory cytokines IFN-gamma and TNF-gamma. Inflammation builds as astrocytes and microglia begin making their own inflammatory cytokines (IL-12, IL-13, IL-23, NO) and the excitotoxic neurotransmitter glutamate.

The next phase of the immune response continues when specific, activated T cells (cytotoxic T) — normally not allowed access to myelin — begin to attack the myelin. The body responds by upregulating the ECS, increasing CB2 receptor expression on astrocytes and microglia and making them more responsive to the inflammatory dampening effects of endocannabinoids. Circulating levels of AEA, OEA, and POA increase as well.

Another mechanism of healing is the activation of CB1 receptors. Agonists of the CB1 receptor work in myriad ways on the T cells themselves, including:

◆ Decreasing production of inflammatory cytokines (IL-2, IFN, TNF-alpha, IL-12)

◆ Decreasing Th1 ability to recognize the myelin as foreign

◆ Decreasing Th1 infiltration to the area

◆ Increasing Th2 production of anti-inflammatory cytokines

Agonists of CB2 receptors induce apoptosis of microglia and helper T cells, decrease TNF-alpha, decrease NO production, decrease IFN-gamma, and increase IL-6. CB2 agonists decrease the replenishment of microglia at the bone marrow, where they are made.

The net effect of cannabinoid receptor binding is a decrease in inflammation, decreased functioning of the T cells causing damage, and a reduction of activated microglia.

Research on Effectiveness

The National Academies of Sciences, Engineering, and Medicine published a report in 2017 citing conclusive evidence for the use of cannabinoids in improving spasticity in multiple sclerosis.

A survey of 112 cannabis-self-medicating people in the United States and United Kingdom found improved symptoms for spasticity, pain, tremor, and depression in more than 90 percent of cases. One study of 37 people utilizing smoked cannabis for three days found relief for pain in multiple sclerosis. A study of 279 individuals for 12 weeks utilizing oral cannabis and Sativex reported relief from muscle stiffness.

A large child/adolescent anxiety multimodal study (CAMS) showed decreased pain and increased mobility over 15 weeks at 2.5 to 5.0 mg THC. Numerous Sativex studies have shown cannabis to be effective at lowering spasticity and sleep disturbances as well as providing relief from pain and bladder incontinence; Sativex is now approved for use in Europe and Australia for these symptoms.

Dosing

Low doses of THC were not effective. It wasn't until THC levels increased above 7.5 mg did people see positive results. I recommend 1:1 MED, with the possible need to build tolerance in order to work up to at least a 7.5 mg dose.

CYTOKINE FUNCTIONS SPECIFIC TO IMMUNE CELLS

IL-23 maintains T cells.

IL-12 causes proliferation of Th1 cells.

IL-10 is anti-inflammatory.

IL-6 potentiates nerve growth factor and decreases TNF-alpha.

IL-4 induces differentiation and activation of Th2 cells.

IL-2 causes proliferation of Th1, B, and natural killer cells.

TGF-beta Th2 supports cell growth and differentiation.

THE
WISDOM
OF CANNABIS

Cannabis and Neural Grooves

Whenever I shoot a free throw while playing basketball, I line up my right toe with a nail in the floorboard. Arrange the ball just so in my hands. Spin the ball toward me. Bend my knees. Dribble twice. Set the ball with the threads across my fingers. Take a deep breath. Look at the front of the rim. Shoot.

I have done this action thousands of times in my basketball career, spent hundreds of hours on the court practicing this very routine. It is so ingrained in my body and mind that now, 32 years since I've played competitively, I still do this routine when I step up to the foul line. This "neural groove" has served me well. My body remembers what to do when I'm exhausted, stressed, overwhelmed, or nervous. My routine actually calms my mind. This is a simple example of a neural groove that has been laid down with practice and, once there, is reflexive without conscious thought. The neural pathway is further ingrained by the reward I receive when the shot goes in: Yay!

Neural grooves are formed for all actions we perform without conscious thought. Physical activities, mental activities, and emotional reactions all use this adaptive strategy. They serve us well, especially in survival situations when, if we were to think about every step, we'd be eaten. They do not serve us well when they are triggered by false information, or if the grooves were laid down when we were very young and had limited ability to respond properly. (Think of all the disagreements we could avoid if we could clear some of our emotional and psychological triggers and their reflexive reactions!) As we age and hopefully develop new skills, our well-worn neural groove reactions may actually cause harm to us and those we love.

Herein lies one of the final gifts of cannabis: help with undoing these unhealthy neural grooves. Perhaps within the neurochemistry of safety cannabis provides, we can begin to unpack the causes of our neural ruts. Perhaps we can get out of them and move into a freer, kinder and more empowered place.

Cannabis Speaks

I am able to create a web of safety within you. A chemical and electrical environment dedicated to bringing you into present time, free of your accumulated baggage that came with learned reflexive behaviors you believed worked to keep you safe. I am the chemistry of safety. I carry with me the neurochemicals designed to have you feel safe.

Your reflexive behaviors have served a purpose to this point. Are you tired of how they isolate you? How they don't offer true safety? How can you be free if you react with a reflex? Reflexes by definition are below conscious awareness. Oh, dear one, you learned to react and protect at a time or situation where you did not have the wisdom or understanding you now have. Isn't it time to utilize some of your hard-won accumulated wisdom right here, right now, when you are emotionally triggered? Triggered to respond in your reflexive, not-so-helpful way? Yes, when you created this pathway you really believed your life depended on it. (Depending on your age and your caregivers, it might be true.) Is it true now? Is it possible you are being robbed of your wisdom by reactions you created when you were not so resourced?

With all of these years of practicing your reaction you have created a quick and decisive neural-grooved reaction. That's what they are designed for, saving you from a severe threat without thinking.

"Line up my toe on the nail. . . . Spin the ball."

You are an expert of your reaction. You don't even need to think about it and *bam*! You are carried along the reaction wave safe from danger.

Is the look your friend gave you actually life threatening?

Mostly that survival response is no longer needed and honestly, a bit melodramatic.

(Sometimes she can be quite funny.)

My gift comes and creates the possibility of you looking at your whole life in present time. I allow you to drop the accumulated energetics of your reactions to respond to what's happening right here. Right now. The sacred pause. I am training wheels for a present time reaction. Eventually you must learn to ride without me. I can offer curiosity about the situation and how it might be different. And as you feel safer, maybe even practicing doing something different in the situation? Create a new neural groove.

"Line up toe, spin spin, dribble dribble, breathe, shoot."

What is possible for healing when you feel safe enough to look at the happenings of life undefended by your accumulated reactive self? Pause. Curiosity. Lightheartedness. Tenderness. Awe. Compassion. Change.

I am medicine for the people to do their internal work. I create the environment of safety for you to experience life with all of your accumulated wisdom right now. I offer this gift freely for you to do your part. In the beginning I might get your attention by altering your perceptions so much you need to stop everything you are doing and pay attention.

Right now!

As you become accustomed to my ways, I won't need to get your attention in the same way. You won't need to be altered so dramatically to take the time to pay attention. As you do your internal work and practice your new neural grooves without me altering your perceptions, you will need my training wheels less and less. Your lack of understanding about this methodology might leave you feeling you should take more so you feel the same level of altered. You don't need it. Your part of our relationship is to practice what you learn with me on your own, without my training wheels.

Think of it as building up a very small muscle.

It takes time and use to become strong.

And you will.

Final Thoughts

We have arrived at the end of our journey together. Perhaps cannabis will resonate with you; perhaps not. The deeper spirit teachings belong to you and cannabis and the relationship you create together, not from something you read in a book.

Ally relationships are not for everyone. Ally relationships with master plants are even fewer. This is the most intimate and committed of relationships. Are you prepared to offer that commitment? To honor, grow, make medicine, spend time journeying, do the work you need to do to be in right relationship?

The answer can be no. As we say on the basketball court, "no harm, no foul." The goal of this book is not to get you to pledge an undying commitment to cannabis. But I would be remiss and disrespectful of *my own* relationship with cannabis if I did not include these final thoughts about allies and neural grooves.

Humans learn best with all of our senses engaged, with our whole being — physical, mental, emotional, and spiritual. When we enter into a sacred ally relationship with cannabis, we must come ready to engage all of our senses and all of our being. We must come ready to be truly known by our ally. Cannabis as an ally can hold us in the most amazing container of safety we have ever known while we do our own internal work.

My lessons from cannabis may be different from yours. We all relate to each other differently; why would it be different with a plant? I hope that sharing my lessons has offered something helpful for you.

Why has cannabis emerged into mainstream consciousness right here and right now? We started our time together with this question, and we end with a few possible answers.

Perhaps she has emerged to help us as individuals and as a collective to stand in present time together, with access to all of our accumulated wisdom. To help us remember who we are.

We are grounded. We are steadfast. We are a force to be reckoned with.

We are safe enough to move and think with our hearts.

Perhaps because cannabis is a master plant and can capture our attention, she is here to remind and teach us how to live in ways we've forgotten.

In right relationship.

With her.

With each other.

With the world around us.

BIBLIOGRAPHY

Chapter 2: Meet the Cannabis Plant

Aizpurua-Olaizola, O., et al. "Evolution of the Cannabinoid and Terpene Content During the Growth of Cannabis Sativa Plants from Different Chemotypes." J Nat Prod 79.2 (2016): 324–31. Print.

Andre, C. M., J. F. Hausman, and G. Guerriero. "Cannabis Sativa: The Plant of the Thousand and One Molecules." Front Plant Sci 7 (2016): 19. Print.

Bartels, E. M., J. Swaddling, and A. P. Harrison. "An Ancient Greek Pain Remedy for Athletes." Pain Pract 6.3 (2006): 212–8. Print.

Chandra, S., et al. "Cannabis Cultivation: Methodological Issues for Obtaining Medical-Grade Product." Epilepsy Behav 70.Pt B (2017): 302–12. Print.

Citti, C., et al. "Analysis of Cannabinoids in Commercial Hemp Seed Oil and Decarboxylation Kinetics Studies of Cannabidiolic Acid (CBDA)." J Pharm Biomed Anal 149 (2018): 532–40. Print.

Clarke, Robert, and Mark Merlin. Cannabis: Evolution and Ethnobotany. University of California Press, first edition, 2016. Print.

Coke, C. J., et al. "Simultaneous Activation of Induced Heterodimerization between Cxcr4 Chemokine Receptor and Cannabinoid Receptor 2 (CB2) Reveals a Mechanism for Regulation of Tumor Progression." J Biol Chem 291.19 (2016): 9991-10005. Print.

Committee on the Health Effects of Marijuana. The Health Effects of Cannabis and Cannabinoids: The Current State of Evidence and Recommendations for Research. Washington, DC: National Academies Press, 2017. Print.

de Meijer, E. P., et al. "The Inheritance of Chemical Phenotype in Cannabis Sativa L." Genetics 163.1 (2003): 335–46. Print.

Deutsch, D. G. "A Personal Retrospective: Elevating Anandamide (AEA) by Targeting Fatty Acid Amide Hydrolase (FAAH) and the Fatty Acid Binding Proteins (FABPS)." Front Pharmacol 7 (2016): 370. Print.

Fujita, W., I. Gomes, and L. A. Devi. "Revolution in GPCR Signalling: Opioid Receptor Heteromers as Novel Therapeutic Targets: Iuphar Review 10." Br J Pharmacol 171.18 (2014): 4155–76. Print.

Gertsch, J., R. G. Pertwee, and V. Di Marzo. "Phytocannabinoids Beyond the Cannabis Plant — Do They Exist?" Br J Pharmacol 160.3 (2010): 523–9. Print.

Grotenhermen, F. "Pharmacokinetics and Pharmacodynamics of Cannabinoids." Clin Pharmacokinet 42.4 (2003): 327–60. Print.

Grotenhermen, F., and K. Muller-Vahl. "The Therapeutic Potential of Cannabis and Cannabinoids." Dtsch Arztebl Int 109.29-30 (2012): 495–501. Print.

Guy, G. W., and P. J. Robson. "A Phase I, Double Blind, Three-Way Crossover Study to Assess the Pharmacokinetic Profile of Cannabis Based Medicine Extract (CBME) Administered Sublingually in Variant Cannabinoid Ratios in Normal Healthy Male Volunteers (GWPK0215)." Journal of Cannabis Therapeutics (The Haworth Integrative Healing Press, an imprint of The Haworth Press, Inc.) 3.4 (2003): 121–52. Print.

Guy, G. W., and P. J. Robson. "A Phase I, Open Label, Four-Way Crossover Study to Compare the Pharmacokinetic Profiles of a Single Dose of 20 Mg of a Cannabis Based Medicine Extract (CBME) Administered on 3 Different Areas of the Buccal Mucosa and to Investigate the Pharmacokinetics of CBME Per Oral in Healthy Male and Female Volunteers (GWPK0112)." Journal of Cannabis Therapeutics (The Haworth Integrative Healing Press, an imprint of The Haworth Press, Inc.) 3.4 (2003): 79–120. Print.

Guy, G. W., and J. M. McPartland. "Models of Cannabis Taxonomy, Cultural Bias, and Conflicts between Scientific and Vernacular Names." The Botanical Review 83.1 (2017). Print.

Hebert-Chatelain, E., et al. "A Cannabinoid Link between Mitochondria and Memory." Nature 539.7630 (2016): 555–59. Print.

Huestis, M. A. "Human Cannabinoid Pharmacokinetics." Chem Biodivers 4.8 (2007): 1770–804. Print.

Karschner, E. L., et al. "Plasma Cannabinoid Pharmacokinetics Following Controlled Oral Delta9-Tetrahydrocannabinol and Oromucosal Cannabis Extract Administration." Clin Chem 57.1 (2011): 66–75. Print.

Lee, S. H., et al. "Multiple Forms of Endocannabinoid and Endovanilloid Signaling Regulate the Tonic Control of Gaba Release." J Neurosci 35.27 (2015): 10039–57. Print.

Lewis, M. A., E. B. Russo, and K. M. Smith. "Pharmacological Foundations of Cannabis Chemovars." Planta Med 84.4 (2018): 225–33. Print.

Massi, P., et al. "Cannabidiol as Potential Anticancer Drug." Br J Clin Pharmacol 75.2 (2013): 303–12. Print.

McPartland, J. M. "Cannabis Systematics at the Levels of Family, Genus, and Species." Cannabis Cannabinoid Res 3.1 (2018): 203–12. Print.

Mead, A. "The Legal Status of Cannabis (Marijuana) and Cannabidiol (CBD) under U.S. Law." Epilepsy Behav 70.Pt B (2017): 288–91. Print.

Morales, P., D. P. Hurst, and P. H. Reggio. "Molecular Targets of the Phytocannabinoids: A Complex Picture." Prog Chem Org Nat Prod 103 (2017): 103–31. Print.

Moreno, E., et al. "Singular Location and Signaling Profile of Adenosine A_2A-Cannabinoid CB_1 Receptor Heteromers in the Dorsal Striatum." Neuropsychopharmacology 43.5 (2018): 964–77. Print.

Moreno-Sanz, G. "Can You Pass the Acid Test? Critical Review and Novel Therapeutic Perspectives of Delta(9)-Tetrahydrocannabinolic Acid A." Cannabis Cannabinoid Res 1.1 (2016): 124–30. Print.

Ohlsson, A., et al. "Plasma Delta-9 Tetrahydrocannabinol Concentrations and Clinical Effects after Oral and Intravenous Administration and Smoking." Clin Pharmacol Ther 28.3 (1980): 409–16. Print.

Oldfield, E., and F. Y. Lin. "Terpene Biosynthesis: Modularity Rules." Angew Chem Int Ed Engl 51.5 (2012): 1124–37. Print.

Pertwee, R. G. "The Diverse CB1 and CB2 Receptor Pharmacology of Three Plant Cannabinoids: Delta9-Tetrahydrocannabinol, Cannabidiol and Delta9-Tetrahydrocannabivarin." Br J Pharmacol 153.2 (2008): 199–215. Print.

Pertwee, R. G. Handbook of Cannabis. Oxford, UK: Oxford University Press, 2014. Print.

Przybyla, J. A., and V. J. Watts. "Ligand-Induced Regulation and Localization of Cannabinoid CB1 and Dopamine D2L Receptor Heterodimers." J Pharmacol Exp Ther 332.3 (2010): 710–9. Print.

Russo, E. B. "Cannabis Roots: A Neglected Herbal Resource." Medicinal Cannabis Conference 2016, Arcata Community Center, Arcata, California, April 2016. Lecture.

Russo, E. B. "Taming THC: Potential Cannabis Synergy and Phytocannabinoid-Terpenoid Entourage Effects." Br J Pharmacol 163.7 (2011): 1344–64. Print.

Russo, E. B., and G. W. Guy. "A Tale of Two Cannabinoids: The Therapeutic Rationale for Combining Tetrahydrocannabinol and Cannabidiol." Med Hypotheses 66.2 (2006): 234–46. Print.

Russo, E. B., and J. Marcu. "Cannabis Pharmacology: The Usual Suspects and a Few Promising Leads." Adv Pharmacol 80 (2017): 67–134. Print.

Singh, M., D. Mamania, and V. Shinde. "The Scope of Hemp (Cannabis sativa L.) Use in Historical Conservation in India." Indian Journal of Traditional Knowledge 17.2 (2018): 314–21. Print.

Small, E. "Evolution and Classification of Cannabis Sativa (Marijuana, Hemp) in Relation to Human Utilization." (2015). Print.

Stinchcomb, A. L., et al. "Human Skin Permeation of Delta8-Tetrahydrocannabinol, Cannabidiol and Cannabinol." J Pharm Pharmacol 56.3 (2004): 291–7. Print.

Szaflarski, M., and J. I. Sirven. "Social Factors in Marijuana Use for Medical and Recreational Purposes." Epilepsy Behav 70.Pt B (2017): 280–87. Print.

Valiveti, S., et al. "In Vitro/in Vivo Correlation Studies for Transdermal Delta 8-THC Development." J Pharm Sci 93.5 (2004): 1154–64. Print.

Wager-Miller, J., R. Westenbroek, and K. Mackie. "Dimerization of G Protein-Coupled Receptors: CB1 Cannabinoid Receptors as an Example." Chem Phys Lipids 121.1-2 (2002): 83–9. Print.

Ward, R. J., J. D. Pediani, and G. Milligan. "Heteromultimerization of Cannabinoid CB(1) Receptor and Orexin OX(1) Receptor Generates a Unique Complex in Which Both Protomers Are Regulated by Orexin A." J Biol Chem 286.43 (2011): 37414–28. Print.

Weiblen, G. D., et al. "Gene Duplication and Divergence Affecting Drug Content in Cannabis Sativa." New Phytol 208.4 (2015): 1241–50. Print.

Zaami, S., et al. "Medical Use of Cannabis: Italian and European Legislation." Eur Rev Med Pharmacol Sci 22.4 (2018): 1161–67. Print.

Chapter 3:
The Endocannabinoid System

Aizpurua-Olaizola, O., et al. "Evolution of the Cannabinoid and Terpene Content During the Growth of Cannabis Sativa Plants from Different Chemotypes." J Nat Prod 79.2 (2016): 324–31. Print.

Anand, P., et al. "Targeting CB2 Receptors and the Endocannabinoid System for the Treatment of Pain." Brain Res Rev 60.1 (2009): 255–66. Print.

Andre, C. M., J. F. Hausman, and G. Guerriero. "Cannabis Sativa: The Plant of the Thousand and One Molecules." Front Plant Sci 7 (2016): 19. Print.

Bab, I., and A. Zimmer. "Cannabinoid Receptors and the Regulation of Bone Mass." Br J Pharmacol 153.2 (2008): 182–8. Print.

Bambico, F. R., et al. "The Fatty Acid Amide Hydrolase Inhibitor URB597 Modulates Serotonin-Dependent Emotional Behaviour, and Serotonin1A and Serotonin2A/C Activity in the Hippocampus." Eur Neuropsychopharmacol 26.3 (2016): 578–90. Print.

Bermudez-Silva, F. J., et al. "The Endocannabinoid System, Eating Behavior and Energy Homeostasis: The End or a New Beginning?" Pharmacol Biochem Behav 95.4 (2010): 375–82. Print.

Bloomfield, M. A., et al. "The Effects of Delta(9)-Tetrahydrocannabinol on the Dopamine System." Nature 539.7629 (2016): 369–77. Print.

Booz, G. W. "Cannabidiol as an Emergent Therapeutic Strategy for Lessening the Impact of Inflammation on Oxidative Stress." Free Radic Biol Med 51.5 (2011): 1054–61. Print.

Braida, D., et al. "5-HT1A Receptors Are Involved in the Anxiolytic Effect of Delta9-Tetrahydrocannabinol and AM 404, the Anandamide Transport Inhibitor, in Sprague-Dawley Rats." Eur J Pharmacol 555.2-3 (2007): 156–63. Print.

Buckner, R. L., J. R. Andrews-Hanna, and D. L. Schacter. "The Brain's Default Network: Anatomy, Function, and Relevance to Disease." Ann N Y Acad Sci 1124 (2008): 1–38. Print.

Burstein, S. "Cannabidiol (CBD) and Its Analogs: A Review of Their Effects on Inflammation." Bioorg Med Chem 23.7 (2015): 1377–85. Print.

Carlini, E. A., et al. "Effects of Marihuana in Laboratory Animals and in Man." Br J Pharmacol 50.2 (1974): 299–309. Print.

Citti, C., et al. "Analysis of Cannabinoids in Commercial Hemp Seed Oil and Decarboxylation Kinetics Studies of Cannabidiolic Acid (CBDA)." J Pharm Biomed Anal 149 (2018): 532–40. Print.

Coke, C. J., et al. "Simultaneous Activation of Induced Heterodimerization between CXCR4 Chemokine Receptor and Cannabinoid Receptor 2 (CB2) Reveals a Mechanism for Regulation of Tumor Progression." J Biol Chem 291.19 (2016): 9991–10005. Print.

Committee on the Health Effects of Marijuana. The Health Effects of Cannabis and Cannabinoids: The Current State of Evidence and Recommendations for Research. Washington, DC: National Academies Press, 2017. Print.

de Fonseca, F. R., et al. "The Endocannabinoid System: Physiology and Pharmacology." Alcohol and Alcoholism 40.1 (2005). Print.

De Gregorio, D., et al. "Cannabidiol Modulates Serotonergic Transmission and Reverses Both Allodynia and Anxiety-Like Behavior in a Model of Neuropathic Pain." Pain 160.1 (2019): 136–50. Print.

Deutsch, D. G. "A Personal Retrospective: Elevating Anandamide (AEA) by Targeting Fatty Acid Amide Hydrolase (FAAH) and the Fatty Acid Binding Proteins (FABPS)." Front Pharmacol 7 (2016): 370. Print.

Di Tomaso, E; M. Beltramo; D. Piomelli. "Brain Cannabinoids in Chocolate." Nature 382 (1996). Print.

Dincheva, I., et al. "FAAH Genetic Variation Enhances Fronto-Amygdala Function in Mouse and Human." Nat Commun 6 (2015): 6395. Print.

Fairbairn, J. W., and J. T. Pickens. "Activity of Cannabis in Relation to Its Delta'-Trans-Tetrahydro-Cannabinol Content." Br J Pharmacol 72.3 (1981): 401–9. Print.

Fisar, Z. "Inhibition of Monoamine Oxidase Activity by Cannabinoids." Naunyn Schmiedebergs Arch Pharmacol 381.6 (2010): 563–72. Print.

Fogaca, M. V., et al. "The Anxiolytic Effects of Cannabidiol in Chronically Stressed Mice Are Mediated by the Endocannabinoid System: Role of Neurogenesis and Dendritic Remodeling." Neuropharmacology 135 (2018): 22–33. Print.

Fride, E. "The Endocannabinoid-CB(1) Receptor System in Pre- and Postnatal Life." Eur J Pharmacol 500.1-3 (2004): 289–97. Print.

Fujita, W., I. Gomes, and L. A. Devi. "Revolution in GPCR Signalling: Opioid Receptor Heteromers as Novel Therapeutic Targets: Iuphar Review 10." Br J Pharmacol 171.18 (2014): 4155–76. Print.

Galindo, L., et al. "Cannabis Users Show Enhanced Expression of CB1-5Ht2A Receptor Heteromers in Olfactory Neuroepithelium Cells." Mol Neurobiol 55.8 (2018): 6347–61. Print.

Galli, J. A., R. A. Sawaya, and F. K. Friedenberg. "Cannabinoid Hyperemesis Syndrome." Curr Drug Abuse Rev 4.4 (2011): 241–9. Print.

Gertsch, J., R. G. Pertwee, and V. Di Marzo. "Phytocannabinoids Beyond the Cannabis Plant — Do They Exist?" Br J Pharmacol 160.3 (2010): 523–9. Print.

Giacoppo, S., et al. "Cannabinoids: New Promising Agents in the Treatment of Neurological Diseases." Molecules 19.11 (2014): 18781–816. Print.

Grotenhermen, F. "Pharmacokinetics and Pharmacodynamics of Cannabinoids." Clin Pharmacokinet 42.4 (2003): 327–60. Print.

Grotenhermen, F., and K. Muller-Vahl. "The Therapeutic Potential of Cannabis and Cannabinoids." Dtsch Arztebl Int 109.29-30 (2012): 495–501. Print.

Guy, G. W., and P. J. Robson. "A Phase I, Double Blind, Three-Way Crossover Study to Assess the Pharmacokinetic Profile of Cannabis Based Medicine Extract (CBME) Administered Sublingually in Variant Cannabinoid Ratios in Normal Healthy Male Volunteers (GWPK0215)." Journal of Cannabis Therapeutics (The Haworth Integrative Healing Press, an imprint of The Haworth Press, Inc.) 3.4 (2003): 121–52. Print.

Guy, G. W., and P. J. Robson. "A Phase I, Open Label, Four-Way Crossover Study to Compare the Pharmacokinetic Profiles of a Single Dose of 20 Mg of a Cannabis Based Medicine Extract (CBME) Administered on 3 Different Areas of the Buccal Mucosa and to Investigate the Pharmacokinetics of CBME Per Oral in Healthy Male and Female Volunteers (GWPK0112)." Journal of Cannabis Therapeutics (The Haworth Integrative Healing Press, an imprint of The Haworth Press, Inc.) 3.4 (2003): 79–120. Print.

Hebert-Chatelain, E., et al. "A Cannabinoid Link between Mitochondria and Memory." Nature 539.7630 (2016): 555–59. Print.

Russo, E. B., and A. G. Hohmann. "Role of Cannabinoids in Pain Management." Comprehensive Treatment of Chronic Pain by Medical, Interventional, and Integrative Approaches. Ed. (eds.), T. R. Deer et al.: American Academy of Pain Medicine, 2013. 181–97. Print.

Huestis, M. A. "Human Cannabinoid Pharmacokinetics." Chem Biodivers 4.8 (2007): 1770–804. Print.

Izzo, A. A., and M. Camilleri. "Emerging Role of Cannabinoids in Gastrointestinal and Liver Diseases: Basic and Clinical Aspects." Gut 57.8 (2008): 1140–55. Print.

Karschner, E. L., et al. "Plasma Cannabinoid Pharmacokinetics Following Controlled Oral Delta9-Tetrahydrocannabinol and Oromucosal Cannabis Extract Administration." Clin Chem 57.1 (2011): 66–75. Print.

Laprairie, R. B., et al. "Cannabidiol Is a Negative Allosteric Modulator of the Cannabinoid CB1 Receptor." Br J Pharmacol 172.20 (2015): 4790–805. Print.

Lee, S. H., et al. "Multiple Forms of Endocannabinoid and Endovanilloid Signaling Regulate the Tonic Control of Gaba Release." J Neurosci 35.27 (2015): 10039–57. Print.

Lewis, M. A., E. B. Russo, and K. M. Smith. "Pharmacological Foundations of Cannabis Chemovars." Planta Med 84.4 (2018): 225–33. Print.

Mallat, A., et al. "The Endocannabinoid System as a Key Mediator During Liver Diseases: New Insights and Therapeutic Openings." Br J Pharmacol 163.7 (2011): 1432–40. Print.

Malone, D. T., and D. A. Taylor. "Involvement of Somatodendritic 5-HT(1A) Receptors in Delta(9)-Tetrahydrocannabinol-Induced Hypothermia in the Rat." Pharmacol Biochem Behav 69.3-4 (2001): 595–601. Print.

Marzo, Vincenzo Di. "Targeting the Endocannabinoid System: To Enhance or Reduce?" Nature 7 (2008). Print.

Massi, P., et al. "Cannabidiol as Potential Anticancer Drug." Br J Clin Pharmacol 75.2 (2013): 303–12. Print.

McPartland, J. M. "Cannabis and Eicosanoids: A Review of Molecular Pharmacology." Journal of Cannabis Therapeutics 1 (2001). Print.

McPartland, J. M. "Expression of the Endocannabinoid System in Fibroblasts and Myofascial Tissues." J Bodyw Mov Ther 12.2 (2008): 169–82. Print.

McPartland, J. M. "The Endocannabinoid System: An Osteopathic Perspective." J Am Osteopath Assoc 108.10 (2008): 586–600. Print.

McPartland, J. M., et al. "Are Cannabidiol and Delta9-Tetrahydrocannabivarin Negative Modulators of the Endocannabinoid System? A Systematic Review." British Journal of Pharmacology (2015). Print.

McPartland, J. M., G. W. Guy, and V. Di Marzo. "Care and Feeding of the Endocannabinoid System: A Systematic Review of Potential Clinical Interventions That Upregulate the Endocannabinoid System." PLoS One 9.3 (2014): e89566. Print.

Mechoulam, R., and L. A. Parker. "The Endocannabinoid System and the Brain." Annu Rev Psychol 64 (2013): 21–47. Print.

Montecucco, F., and V. Di Marzo. "At the Heart of the Matter: The Endocannabinoid System in Cardiovascular Function and Dysfunction." Trends Pharmacol Sci 33.6 (2012): 331–40. Print.

Morales, P., D. P. Hurst, and P. H. Reggio. "Molecular Targets of the Phytocannabinoids: A Complex Picture." Prog Chem Org Nat Prod 103 (2017): 103–31. Print.

Moreno, E., et al. "Singular Location and Signaling Profile of Adenosine A2A-Cannabinoid CB1 Receptor Heteromers in the Dorsal Striatum." Neuropsychopharmacology 43.5 (2018): 964–77. Print.

Moreno-Sanz, G. "Can You Pass the Acid Test? Critical Review and Novel Therapeutic Perspectives of Delta(9)-Tetrahydrocannabinolic Acid A." Cannabis Cannabinoid Res 1.1 (2016): 124–30. Print.

Ohlsson, A., et al. "Plasma Delta-9 Tetrahydrocannabinol Concentrations and Clinical Effects after Oral and Intravenous Administration and Smoking." Clin Pharmacol Ther 28.3 (1980): 409–16. Print.

Oldfield, E., and F. Y. Lin. "Terpene Biosynthesis: Modularity Rules." Angew Chem Int Ed Engl 51.5 (2012): 1124–37. Print.

Oz, M. "Receptor-Independent Actions of Cannabinoids on Cell Membranes: Focus on Endocannabinoids." Pharmacol Ther 111.1 (2006): 114–44. Print.

Pacher, P., S. Batkai, and G. Kunos. "The Endocannabinoid System as an Emerging Target of Pharmacotherapy." Pharmacol Rev 58.3 (2006): 389–462. Print.

Pacher, P., and G. Kunos. "Modulating the Endocannabinoid System in Human Health and Disease — Successes and Failures." FEBS J 280.9 (2013): 1918–43. Print.

Pacher, P., and R. Mechoulam. "Is Lipid Signaling through Cannabinoid 2 Receptors Part of a Protective System?" Prog Lipid Res 50.2 (2011): 193–211. Print.

Pamplona, F. A., et al. "Anti-Inflammatory Lipoxin A4 Is an Endogenous Allosteric Enhancer of CB1 Cannabinoid Receptor." Proc Natl Acad Sci U S A 109.51 (2012): 21134–9. Print.

Pava, M. J., et al. "Endocannabinoid Modulation of Cortical Up-States and NREM Sleep." PLoS One 9.2 (2014): e88672. Print.

Pertwee, R. G. "Targeting the Endocannabinoid System with Cannabinoid Receptor Agonists: Pharmacological Strategies and Therapeutic Possibilities." Philos Trans R Soc Lond B Biol Sci 367.1607 (2012): 3353–63. Print.

Pertwee, R. G. "The Diverse CB1 and CB2 Receptor Pharmacology of Three Plant Cannabinoids: Delta9-Tetrahydrocannabinol, Cannabidiol and Delta9-Tetrahydrocannabivarin." Br J Pharmacol 153.2 (2008): 199–215. Print.

Pertwee, R. G. "The Therapeutic Potential of Drugs That Target Cannabinoid Receptors or Modulate the Tissue Levels or Actions of Endocannabinoids." AAPS J 7.3 (2005): E625–54. Print.

Pertwee, R. G. Handbook of Cannabis. Oxford, UK: Oxford University Press, 2014. Print.

Przybyla, J. A., and V. J. Watts. "Ligand-Induced Regulation and Localization of Cannabinoid CB1 and Dopamine D2l Receptor Heterodimers." J Pharmacol Exp Ther 332.3 (2010): 710–9. Print.

Ramsay, D. S., and S. C. Woods. "Clarifying the Roles of Homeostasis and Allostasis in Physiological Regulation." Psychol Rev 121.2 (2014): 225–47. Print.

Rock, E. M., et al. "Cannabidiol, a Non-Psychotropic Component of Cannabis, Attenuates Vomiting and Nausea-Like Behaviour Via Indirect Agonism of 5-HT(1A) Somatodendritic Autoreceptors in the Dorsal Raphe Nucleus." Br J Pharmacol 165.8 (2012): 2620–34. Print.

Rubio-Araiz, A., et al. "The Endocannabinoid System Modulates a Transient TNF Pathway That Induces Neural Stem Cell Proliferation." Mol Cell Neurosci 38.3 (2008): 374–80. Print.

Russo, E. B. "Clinical Endocannabinoid Deficiency Reconsidered: Current Research Supports the Theory in Migraine, Fibromyalgia, Irritable Bowel, and Other Treatment-Resistant Syndromes." Cannabis Cannabinoid Res 1.1 (2016): 154–65. Print.

Russo, E. B. "Taming THC: Potential Cannabis Synergy and Phytocannabinoid-Terpenoid Entourage Effects." Br J Pharmacol 163.7 (2011): 1344–64. Print.

Russo, E. B., and G. W. Guy. "A Tale of Two Cannabinoids: The Therapeutic Rationale for Combining Tetrahydrocannabinol and Cannabidiol." Med Hypotheses 66.2 (2006): 234–46. Print.

Sales, A. J., et al. "Antidepressant-Like Effect Induced by Cannabidiol Is Dependent on Brain Serotonin Levels." Prog Neuropsycho-pharmacol Biol Psychiatry 86 (2018): 255–61. Print.

Sartim, A. G., F. S. Guimaraes, and S. R. Joca. "Antidepressant-Like Effect of Cannabidiol Injection into the Ventral Medial Prefrontal Cortex-Possible Involvement of 5-HT1A and CB1 Receptors." Behav Brain Res 303 (2016): 218–27. Print.

Scott, K. A., A. G. Dalgleish, and W. M. Liu. "The Combination of Cannabidiol and Delta9-Tetrahydrocannabinol Enhances the Anticancer Effects of Radiation in an Orthotopic Murine Glioma Model." Mol Cancer Ther 13.12 (2014): 2955–67. Print.

Seyrek, M., et al. "Systemic Cannabinoids Produce CB(1)-Mediated Antinociception by Activation of Descending Serotonergic Pathways That Act Upon Spinal 5-HT(7) and 5-HT(2A) Receptors." Eur J Pharmacol 649.1-3 (2010): 183–94. Print.

Smirnov, M. S., and E. A. Kiyatkin. "Behavioral and Temperature Effects of Delta 9-Tetrahydrocannabinol in Human-Relevant Doses in Rats." Brain Res 1228 (2008): 145–60. Print.

Smith, S. C., and M. S. Wagner. "Clinical Endocannabinoid Deficiency (CECD) Revisited: Can This Concept Explain the Therapeutic Benefits of Cannabis in Migraine, Fibromyalgia, Irritable Bowel Syndrome and Other Treatment-Resistant Conditions?" Neuro Endocrinol Lett 35.3 (2014): 198–201. Print.

Stinchcomb, A. L., et al. "Human Skin Permeation of Delta8-Tetrahydrocannabinol, Cannabidiol and Cannabinol." J Pharm Pharmacol 56.3 (2004): 291–7. Print.

Szkudlarek, H. J., et al. "Delta-9-Tetrahydrocannabinol and Cannabidiol Produce Dissociable Effects on Prefrontal Cortical Executive Function and Regulation of Affective Behaviors." Neuropsychopharmacology 44.4 (2019): 817–25. Print.

Valiveti, S., et al. "In Vitro/in Vivo Correlation Studies for Transdermal Delta 8-THC Development." J Pharm Sci 93.5 (2004): 1154–64. Print.

Velenovska, M., and Z. Fisar. "Effect of Cannabinoids on Platelet Serotonin Uptake." Addict Biol 12.2 (2007): 158–66. Print.

Vinals, X., et al. "Cognitive Impairment Induced by Delta9-Tetrahydrocannabinol Occurs through Heteromers between Cannabinoid CB1 and Serotonin 5-HT2A Receptors." PLoS Biol 13.7 (2015): e1002194. Print.

Wager-Miller, J., R. Westenbroek, and K. Mackie. "Dimerization of G Protein-Coupled Receptors: CB1 Cannabinoid Receptors as an Example." Chem Phys Lipids 121.1-2 (2002): 83–9. Print.

Ward, R. J., J. D. Pediani, and G. Milligan. "Heteromultimerization of Cannabinoid CB(1) Receptor and Orexin OX(1) Receptor Generates a Unique Complex in Which Both Protomers Are Regulated by Orexin A." J Biol Chem 286.43 (2011): 37414–28. Print.

Wilkinson, J. D., et al. "Medicinal Cannabis: Is Delta9-Tetrahydrocannabinol Necessary for All Its Effects?" J Pharm Pharmacol 55.12 (2003): 1687–94. Print.

Zhornitsky, S., and S. Potvin. "Cannabidiol in Humans — the Quest for Therapeutic Targets." Pharmaceuticals (Basel) 5.5 (2012): 529–52. Print.

Chapter 4: Preparation and Dosage

Bonn-Miller, M. O., et al. "Labeling Accuracy of Cannabidiol Extracts Sold Online." JAMA 318.17 (2017): 1708–09. Print.

Casiraghi, A., et al. "Extraction Method and Analysis of Cannabinoids in Cannabis Olive Oil Preparations." Planta Med 84.4 (2018): 242–49. Print.

Committee on the Health Effects of Marijuana. The Health Effects of Cannabis and Cannabinoids: The Current State of Evidence and Recommendations for Research. Washington, DC: National Academies Press, 2017. Print.

Fiorini, D., et al. "Valorizing Industrial Hemp (Cannabis Sativa L.) By-Products: Cannabidiol Enrichment in the Inflorescence Essential Oil Optimizing Sample Pre-Treatment Prior to Distillation." Industrial Crops and Products 128 (2019). Print.

Hazekamp, A., et al. "Cannabis Tea Revisited: A Systematic Evaluation of the Cannabinoid Composition of Cannabis Tea." J Ethnopharmacol 113.1 (2007): 85–90. Print.

Hosseini, A., et al. "Starting Dose Calculation for Medicinal Plants in Animal Studies; Recommendation of a Simple and Reliable Method." Research Journal of Pharmacognosy 5.2 (2018). Print.

Jikomes, N., and M. Zoorob. "The Cannabinoid Content of Legal Cannabis in Washington State Varies Systematically across Testing Facilities and Popular Consumer Products." Sci Rep 8.1 (2018): 4519. Print.

Lanz, C., et al. "Medicinal Cannabis: In Vitro Validation of Vaporizers for the Smoke-Free Inhalation of Cannabis." PLoS One 11.1 (2016): e0147286. Print.

Lin, T. K., L. Zhong, and J. L. Santiago. "Anti-Inflammatory and Skin Barrier Repair Effects of Topical Application of Some Plant Oils." Int J Mol Sci 19.1 (2017). Print.

Lindholst, Christian. "Long Term Stability of Cannabis Resin and Cannabis Extracts." Australian Journal of Forensic Sciences 42.3 (2010): 181–90. Print.

MacCallum, C. A., and E. B. Russo. "Practical Considerations in Medical Cannabis Administration and Dosing." Eur J Intern Med 49 (2018): 12–19. Print.

McPartland, J. M., and E. B. Russo. "Cannabis and Cannabis Extracts: Greater Than the Sum of Their Parts?" Journal of Cannabis Therapeutics 1.3/4 (2001). Print.

Narayanaswami, K., et al. "Stability of Cannabis Sativa L. Samples and Their Extracts, on Prolonged Storage in Delhi." Bull Narc 30.4 (1978): 57–69. Print.

Oh, D. A., et al. "Effect of Food on the Pharmacokinetics of Dronabinol Oral Solution versus Dronabinol Capsules in Healthy Volunteers." Clin Pharmacol 9 (2017): 9–17. Print.

Ohlsson, A., et al. "Plasma Delta-9 Tetrahydrocannabinol Concentrations and Clinical Effects after Oral and Intravenous Administration and Smoking." Clin Pharmacol Ther 28.3 (1980): 409–16. Print.

Parikh, N., et al. "Bioavailability Study of Dronabinol Oral Solution versus Dronabinol Capsules in Healthy Volunteers." Clin Pharmacol 8 (2016): 155–62. Print.

Pertwee, R. G. Handbook of Cannabis. Oxford, UK: Oxford University Press, 2014. Print.

Peschel, W. "Quality Control of Traditional Cannabis Tinctures: Pattern, Markers, and Stability." Sci Pharm 84.3 (2016): 567–84. Print.

Romano, L. L., and A. Hazekamp. "Cannabis Oil: Chemical Evaluation of an Upcoming Cannabis-Based Medicine." Cannabinoids 7.1 (2013). Print.

Spelman, K. "Home Extractions of Cannabis: Efficiency of Cannabinoid Yield." The Healing Power of Cannabis: Medicinal Uses, Preparation and Organic Cultivation 2018. Lecture notes.

Turner, C. E., et al. "Constituents of Cannabis Sativa L. IV. Stability of Cannabinoids in Stored Plant Material." J Pharm Sci 62.10 (1973): 1601–5. Print.

Vandrey, R., et al. "Cannabinoid Dose and Label Accuracy in Edible Medical Cannabis Products." JAMA 313.24 (2015): 2491–3. Print.

Victory, K. R., et al. "Notes from the Field: Occupational Hazards Associated with Harvesting and Processing Cannabis — Washington, 2015–2016." MMWR Morb Mortal Wkly Rep 67.8 (2018): 259–60. Print.

Wang, M., et al. "Decarboxylation Study of Acidic Cannabinoids: A Novel Approach Using Ultra-High-Performance Supercritical Fluid Chromatography/Photodiode Array-Mass Spectrometry." Cannabis Cannabinoid Res 1.1 (2016): 262–71. Print.

Chapter 5: Contraindications and Considerations

Bidwell, L. C., et al. "A Novel Observational Method for Assessing Acute Responses to Cannabis: Preliminary Validation Using Legal Market Strains." Cannabis Cannabinoid Res 3.1 (2018): 35–44. Print.

Colizzi, M., and R. Murray. "Cannabis and Psychosis: What Do We Know and What Should We Do?" Br J Psychiatry 212.4 (2018): 195–96. Print.

Committee on the Health Effects of Marijuana. "Health Effects of Cannabis and Cannabinoids: The Current State of Evidence and Recommendations for Research." Washington, DC: National Academies Press, 2017. Print.

D'Souza, D. C., et al. "Blunted Psychotomimetic and Amnestic Effects of Delta-9-Tetrahydrocannabinol in Frequent Users of Cannabis." Neuropsychopharmacology 33.10 (2008): 2505–16. Print.

Gani, R., and Z. A. Bhat. "Anxiety Disorders and Herbal Medicines." International Journal of Pharmaceutical Sciences and Research 9.3 (2018). Print.

Greer, G. R., C. S. Grob, and A. L. Halberstadt. "PTSD Symptom Reports of Patients Evaluated for the New Mexico Medical Cannabis Program." J Psychoactive Drugs 46.1 (2014): 73–7. Print.

Hirvonen, J., et al. "Reversible and Regionally Selective Downregulation of Brain Cannabinoid CB1 Receptors in Chronic Daily Cannabis Smokers." Mol Psychiatry 17.6 (2012): 642–9. Print.

Jones, J. L., and K. E. Abernathy. "Successful Treatment of Suspected Cannabinoid Hyperemesis Syndrome Using Haloperidol in the Outpatient Setting." Case Rep Psychiatry 2016 (2016): 3614053. Print.

"Marijuana Cannabis Allergy." American Academy of Allergy, Asthma and Immunology (AAAAI) (2019). Web.

Minerbi, A., W. Hauser, and M. A. Fitzcharles. "Medical Cannabis for Older Patients." Drugs Aging 36.1 (2019): 39–51. Print.

Mukamal, K. J., et al. "An Exploratory Prospective Study of Marijuana Use and Mortality Following Acute Myocardial Infarction." Am Heart J 155.3 (2008): 465–70. Print.

Pertwee, R. G. Handbook of Cannabis. Oxford, UK: Oxford University Press, 2014. Print.

Schmid, K., et al. "The Effects of Cannabis on Heart Rate Variability and Well-Being in Young Men." Pharmacopsychiatry 43.4 (2010): 147–50. Print.

Simonetto, D. A., et al. "Cannabinoid Hyperemesis: A Case Series of 98 Patients." Mayo Clin Proc 87.2 (2012): 114–9. Print.

Solowij, N. "Peering through the Haze of Smoked vs Vaporized Cannabis-to Vape or Not to Vape?" JAMA Netw Open 1.7 (2018): e184838. Print.

Valenti, D. "Marijuana: Impairment to the Visual Sensory System." Innovation in Aging 2 (Suppl 1) (2018). Print.

Walsh, Z., et al. "Medical Cannabis and Mental Health: A Guided Systematic Review." Clin Psychol Rev 51 (2017): 15–29. Print.

Zehra, A., et al. "Cannabis Addiction and the Brain: A Review." J Neuroimmune Pharmacol (2018). Print.

Chapter 6: Conditions and Clinical Applications

Adejumo, A. C., et al. "Cannabis Use Is Associated with Reduced Prevalence of Progressive Stages of Alcoholic Liver Disease." Liver Int 38.8 (2018): 1475–86. Print.

Alfulaij, N., et al. "Cannabinoids, the Heart of the Matter." J Am Heart Assoc 7.14 (2018). Print.

Ali, S., I. E. Scheffer, and L. G. Sadleir. "Efficacy of Cannabinoids in Paediatric Epilepsy." Dev Med Child Neurol (2018). Print.

Allan, G. M., et al. "Systematic Review of Systematic Reviews for Medical Cannabinoids: Pain, Nausea and Vomiting, Spasticity, and Harms. " Can Fam Physician 64.2 (2018): e78–e94. Print.

Amato, L., et al. "Systematic Review of Safeness and Therapeutic Efficacy of Cannabis in Patients with Multiple Sclerosis, Neuropathic Pain, and in Oncological Patients Treated with Chemotherapy." Epidemiol Prev 41.5-6 (2017): 279–93. Print.

Anderson, C. L., et al. "Cannabidiol for the Treatment of Drug-Resistant Epilepsy in Children: Current State of Research." Journal of Pediatric Neurology 14.4 (2017). Print.

Araki, Nobuo. "Migraine." Japan Medical Association Journal 47.3 (2004). Print.

Aran, A., et al. "Brief Report: Cannabidiol-Rich Cannabis in Children with Autism Spectrum Disorder and Severe Behavioral Problems — a Retrospective Feasibility Study." J Autism Dev Disord 49.3 (2019): 1284–88. Print.

Barker, Jonathan. "Review of the Public Health Risks of Widespread Cannabis Use." Rhode Island Medical Journal (2018). Print.

Bar-Lev Schleider, L., et al. "Prospective Analysis of Safety and Efficacy of Medical Cannabis in Large Unselected Population of Patients with Cancer." Eur J Intern Med 49 (2018): 37–43. Print.

Benevenuto, S. G., et al. "Recreational Use of Marijuana During Pregnancy and Negative Gestational and Fetal Outcomes: An Experimental Study in Mice." Toxicology (2017): 94–101. Print.

Bergamaschi, M. M., et al. "Cannabidiol Reduces the Anxiety Induced by Simulated Public Speaking in Treatment-Naive Social Phobia Patients." Neuropsychopharmacology 36.6 (2011): 1219–26. Print.

Bhattacharyya, S., et al. "Acute Induction of Anxiety in Humans by Delta-9-Tetrahydrocannabinol Related to Amygdalar Cannabinoid-1 (CB1) Receptors." Sci Rep 7.1 (2017): 15025. Print.

Blazquez, C., et al. "Loss of Striatal Type 1 Cannabinoid Receptors Is a Key Pathogenic Factor in Huntington's Disease." Brain 134.Pt 1 (2011): 119–36. Print.

Bloomfield, M. A. P., et al. "The Neuropsycho-pharmacology of Cannabis: A Review of Human Imaging Studies." Pharmacol Ther (2018). Print.

Boehnke, K. F., E. Litinas, and D. J. Clauw. "Medical Cannabis Use Is Associated with Decreased Opiate Medication Use in a Retrospective Cross-Sectional Survey of Patients with Chronic Pain." J Pain 17.6 (2016): 739–44. Print.

Bonn-Miller, M. O., et al. "Labeling Accuracy of Cannabidiol Extracts Sold Online." JAMA 318.17 (2017): 1708–09. Print.

Borgelt, L. M., et al. "The Pharmacologic and Clinical Effects of Medical Cannabis." Pharmacotherapy 33.2 (2013): 195–209. Print.

Bradford, A. C., et al. "Association between Us State Medical Cannabis Laws and Opioid Prescribing in the Medicare Part D Population." JAMA Intern Med 178.5 (2018): 667–72. Print.

Brodie, J. S., V. Di Marzo, and G. W. Guy. "Polypharmacology Shakes Hands with Complex Aetiopathology." Trends Pharmacol Sci 36.12 (2015): 802–21. Print.

Brown, M. R. D., and W. P. Farquhar-Smith. "Cannabinoids and Cancer Pain: A New Hope or a False Dawn?" Eur J Intern Med 49 (2018): 30–36. Print.

Buckner, J. D., K. Walukevich Dienst, and M. J. Zvolensky. "Distress Tolerance and Cannabis Craving: The Impact of Laboratory-Induced Distress." Exp Clin Psychopharmacol (2018). Print.

Busquets-Garcia, A., et al. "Peripheral and Central CB1 Cannabinoid Receptors Control Stress-Induced Impairment of Memory Consolidation." Proc Natl Acad Sci U S A 113.35 (2016): 9904–9. Print.

Campos, A. C., et al. "Multiple Mechanisms Involved in the Large-Spectrum Therapeutic Potential of Cannabidiol in Psychiatric Disorders." Philos Trans R Soc Lond B Biol Sci 367.1607 (2012): 3364–78. Print.

Cannabis for Medical Purposes Evidence Guide: Information for Pharmacists and Other Health Care Professionals: Canadian Pharmacists Association, 2018. Print.

Carvalho, A. F., and E. J. Van Bockstaele. "Cannabinoid Modulation of Noradrenergic Circuits: Implications for Psychiatric Disorders." Prog Neuropsychopharmacol Biol Psychiatry 38.1 (2012): 59–67. Print.

Chabarria, K. C., et al. "Marijuana Use and Its Effects in Pregnancy." American Journal of Obstetrics and Gynecology (2016). Print.

Chakrabarti, B., and S. Baron-Cohen. "Variation in the Human Cannabinoid Receptor CNR1 Gene Modulates Gaze Duration for Happy Faces." Mol Autism 2.1 (2011): 10. Print.

Cohen, K., A. Weizman, and A. Weinstein. "Modulatory Effects of Cannabinoids on Brain Neurotransmission." Eur J Neurosci 50.3 (2019): 2322–45. Print.

Committee on the Health Effects of Marijuana. "The Health Effects of Cannabis and Cannabinoids: The Current State of Evidence and Recommendations for Research." Washington, District of Columbia: National Academies Press, 2017. Print.

Conner, S. N., et al. "Maternal Marijuana Use and Neonatal Morbidity." Am J Obstet Gynecol (2015). Print.

Corli, O., et al. "Cannabis as a Medicine. An Update of the Italian Reality." Eur J Intern Med 60 (2019): e9–e10. Print.

Cunha, P., et al. "Endocannabinoid System in Cardiovascular Disorders — New Pharmacotherapeutic Opportunities." J Pharm Bioallied Sci 3.3 (2011): 350–60. Print.

De Aquino, J. P., and D. A. Ross. "Cannabinoids and Pain: Weeding out Undesired Effects with a Novel Approach to Analgesia." Biol Psychiatry 84.10 (2018): e67–e69. Print.

Desroches, J., and P. Beaulieu. "Opioids and Cannabinoids Interactions: Involvement in Pain Management." Curr Drug Targets 11.4 (2010): 462–73. Print.

Devinsky, O., et al. "Cannabidiol: Pharmacology and Potential Therapeutic Role in Epilepsy and Other Neuropsychiatric Disorders." Epilepsia 55.6 (2014): 791–802. Print.

Dhadwal, G., and M. G. Kirchhof. "The Risks and Benefits of Cannabis in the Dermatology Clinic." J Cutan Med Surg 22.2 (2018): 194–99. Print.

Dincheva, I., et al. "Faah Genetic Variation Enhances Fronto-Amygdala Function in Mouse and Human." Nat Commun 6 (2015): 6395. Print.

Dogrul, A., et al. "Topical Cannabinoid Antinociception: Synergy with Spinal Sites." Pain 105.1-2 (2003): 11–6. Print.

Donadelli, M., et al. "Gemcitabine/Cannabinoid Combination Triggers Autophagy in Pancreatic Cancer Cells through a ROS-Mediated Mechanism." Cell Death Dis 2 (2011): e152. Print.

Dong, C., et al. "Cannabinoid Exposure During Pregnancy and Its Impact on Immune Function." Cell Mol Life Sci (2018). Print.

Donvito, G., et al. "The Endogenous Cannabinoid System: A Budding Source of Targets for Treating Inflammatory and Neuropathic Pain." Neuropsychopharmacology 43.1 (2018): 52–79. Print.

Dugas, E. N., et al. "Early Risk Factors for Daily Cannabis Use in Young Adults." Can J Psychiatry (2018): 706743718804541. Print.

Elikkottil, J., P. Gupta, and K. Gupta. "The Analgesic Potential of Cannabinoids." J Opioid Manag 5.6 (2009): 341–57. Print.

El Marroun, H., et al. "Intrauterine Cannabis Exposure Affects Fetal Growth Trajectories: The Generation R Study." J Am Acad Child Adolesc Psychiatry 48.12 (2009): 1173–81. Print.

Engels, F. K., et al. "Medicinal Cannabis Does Not Influence the Clinical Pharmacokinetics of Irinotecan and Docetaxel." Oncologist 12.3 (2007): 291–300. Print.

Ferguson, G., and M. A. Ware. "Review Article: Sleep, Pain and Cannabis." Journal of Sleep Disorders and Therapy 4.2 (2015). Print.

Fine, P. G., and M. J. Rosenfeld. "The Endocannabinoid System, Cannabinoids, and Pain." Rambam Maimonides Med J 4.4 (2013): e0022. Print.

Fowler, C. J., et al. "Targeting the Endocannabinoid System for the Treatment of Cancer — a Practical View." Curr Top Med Chem 10.8 (2010): 814–27. Print.

Fraguas-Sanchez, A. I., and A. I. Torres-Suarez. "Medical Use of Cannabinoids." Drugs 78.16 (2018): 1665–703. Print.

Gaffal, E., et al. "Anti-Inflammatory Activity of Topical THC in DNFB-Mediated Mouse Allergic Contact Dermatitis Independent of CB1 and CB2 Receptors." Allergy 68.8 (2013): 994–1000. Print.

Gallily, R., Z. Yekhtin, and L. O. Hanus. "Overcoming the Bell-Shaped Dose-Response of Cannabidiol by Using Cannabis Extract Enriched in Cannabidiol." Pharmacology and Pharmacy 6 (2015). Print.

Gates, P. J., L. Albertella, and J. Copeland. "The Effects of Cannabinoid Administration on Sleep: A Systematic Review of Human Studies." Sleep Med Rev 18.6 (2014): 477–87. Print.

Giacoppo, S., et al. "A New Formulation of Cannabidiol in Cream Shows Therapeutic Effects in a Mouse Model of Experimental Autoimmune Encephalomyelitis." Daru 23 (2015): 48. Print.

Gobbi, G., et al. "Antidepressant-Like Activity and Modulation of Brain Monoaminergic Transmission by Blockade of Anandamide Hydrolysis." Proc Natl Acad Sci U S A 102.51 (2005): 18620–5. Print.

Gorter, R. "Cannabis in Pain Management." Townsend Letter (2018). Print.

Gunduz-Cinar, O., et al. "Convergent Translational Evidence of a Role for Anandamide in Amygdala-Mediated Fear Extinction, Threat Processing and Stress-Reactivity." Mol Psychiatry 18.7 (2013): 813–23. Print.

Guzman, M., et al. "A Pilot Clinical Study of Delta9-Tetrahydrocannabinol in Patients with Recurrent Glioblastoma Multiforme." Br J Cancer 95.2 (2006): 197–203. Print.

Haroutounian, S., et al. "The Effect of Medicinal Cannabis on Pain and Quality-of-Life Outcomes in Chronic Pain: A Prospective Open-Label Study." Clin J Pain 32.12 (2016): 1036–43. Print.

Hess, C., M. Kramer, and B. Madea. "Topical Application of THC Containing Products Is Not Able to Cause Positive Cannabinoid Finding in Blood or Urine." Forensic Sci Int 272 (2017): 68–71. Print.

Hosseini, S., and M. Oremus. "The Effect of Age of Initiation of Cannabis Use on Psychosis, Depression, and Anxiety among Youth under 25 Years." Can J Psychiatry (2018): 706743718809339. Print.

Huang, Y., et al. "The Role of Traditional Chinese Herbal Medicines and Bioactive Ingredients on Ion Channels: A Brief Review and Prospect." CNS Neurol Disord Drug Targets (2018). Print.

Huestis, M. A. "Human Cannabinoid Pharmacokinetics." Chem Biodivers 4.8 (2007): 1770–804. Print.

Huizink, A. C. "Prenatal Cannabis Exposure and Infant Outcomes: Overview of Studies." Prog Neuropsychopharmacol Biol Psychiatry 52 (2014): 45–52. Print.

Hupli, A. M. M. "Medical Cannabis for Adult Attention Deficit Hyperactivity Disorder: Sociological Patient Case Report of Cannabinoid Therapeutics in Finland." Medical Cannabis and Cannabinoids (2018). Print.

Ibarra-Lecue, I., et al. "The Endocannabinoid System in Mental Disorders: Evidence from Human Brain Studies." Biochem Pharmacol 157 (2018): 97–107. Print.

"International Association for Cannabinoid Medicines Bulletin." 2017. Web.

Javed, H., et al. "Cannabinoid Type 2 (CB2) Receptors Activation Protects against Oxidative Stress and Neuroinflammation Associated Dopaminergic Neurodegeneration in Rotenone Model of Parkinson's Disease." Front Neurosci 10 (2016): 321. Print.

Jean-Gilles, L., et al. "Effects of Pro-Inflammatory Cytokines on Cannabinoid CB1 and CB2 Receptors in Immune Cells." Acta Physiol (Oxf) 214.1 (2015): 63–74. Print.

Jordt, S.-E., et al. "Mustard Oils and Cannabinoids Excite Sensory Nerve Fibres through the TRP Channel ANKTM1." Nature 427 (2004). Print.

Kander, J. Cannabis for the Treatment of Cancer. Ed. Dennis Hill. 2015. The Anticancer Activity of Phytocannabinoids and Endocannabinoids. Web <http://coscc.org/wp-content/uploads/2015/07/Cannabis-and-Cancer.pdf>.

Karhson, D. S., et al. "Plasma Anandamide Concentrations Are Lower in Children with Autism Spectrum Disorder." Mol Autism 9 (2018): 18. Print.

Karl, T., et al. "The Therapeutic Potential of the Endocannabinoid System for Alzheimer's Disease." Expert Opin Ther Targets 16.4 (2012): 407–20. Print.

Karst, M., and S. Wippermann. "Cannabinoids against Pain. Efficacy and Strategies to Reduce Psychoactivity: A Clinical Perspective." Expert Opin Investig Drugs 18.2 (2009): 125–33. Print.

Kinnucan, J. "Use of Medical Cannabis in Patients with Inflammatory Bowel Disease." Gastroenterology and Hepatology (2018). Print.

Kirkedal, C., et al. "Hemisphere-Dependent Endocannabinoid System Activity in Prefrontal Cortex and Hippocampus of the Flinders Sensitive Line Rodent Model of Depression." Neurochem Int 125 (2019): 7–15. Print.

Klein, T. W. "Cannabinoid-Based Drugs as Anti-Inflammatory Therapeutics." Nat Rev Immunol 5.5 (2005): 400–11. Print.

Kunos, G., et al. "Endocannabinoids as Cardio-vascular Modulators." Chem Phys Lipids 108.1-2 (2000): 159–68. Print.

Kupczyk, P., A. Reich, and J. C. Szepietowski. "Cannabinoid System in the Skin — a Possible Target for Future Therapies in Dermatology." Exp Dermatol 18.8 (2009): 669–79. Print.

Ladouceur, R. "The Cannabis Paradox." Can Fam Physician 64.2 (2018): 86. Print.

Landa, L., et al. "Medical Cannabis in the Treatment of Cancer Pain and Spastic Conditions and Options of Drug Delivery in Clinical Practice." Biomed Pap Med Fac Univ Palacky Olomouc Czech Repub 162.1 (2018): 18–25. Print.

Leung, L. "Cannabis and Its Derivatives: Review of Medical Use." J Am Board Fam Med 24.4 (2011): 452–62. Print.

Leweke, F. M., et al. "Cannabidiol Enhances Anandamide Signaling and Alleviates Psychotic Symptoms of Schizophrenia." Transl Psychiatry 2 (2012): e94. Print.

Lim, M., and M. G. Kirchhof. "Dermatology-Related Uses of Medical Cannabis Promoted by Dispensaries in Canada, Europe, and the United States." J Cutan Med Surg 23.2 (2019): 178–84. Print.

Lin, T. K., L. Zhong, and J. L. Santiago. "Anti-Inflammatory and Skin Barrier Repair Effects of Topical Application of Some Plant Oils." Int J Mol Sci 19.1 (2017). Print.

Lodzki, M., et al. "Cannabidiol-Transdermal Delivery and Anti-Inflammatory Effect in a Murine Model." J Control Release 93.3 (2003): 377–87. Print.

Lynch, M. E., M. A. Ware. "Cannabinoids for the Treatment of Chronic Non-Cancer Pain: An Updated Systematic Review of Randomized Controlled Trials." Journal of Neuroimmune Pharmacology (2015). Print.

Mark, K., and M. Terplan. "Cannabis and Pregnancy: Maternal Child Health Implications During a Period of Drug Policy Liberalization." Prev Med 104 (2017): 46–49. Print.

McAllister, S. D., L. Soroceanu, and P. Y. Desprez. "The Antitumor Activity of Plant-Derived Non-Psychoactive Cannabinoids." J Neuroimmune Pharmacol 10.2 (2015): 255–67. Print.

McGuire, P., et al. "Cannabidiol (CBD) as an Adjunctive Therapy in Schizophrenia: A Multicenter Randomized Controlled Trial." Am J Psychiatry 175.3 (2018): 225–31. Print.

McSweeney, L. J., P. McEneaney, and S. O'Reilly. "Cannabis Versus Combination Chemotherapy; N = 1 Trial in Hodgkin's Lymphoma." Ir J Med Sci (2018). Print.

Mendiguren, A., E. Aostri, and J. Pineda. "Regulation of Noradrenergic and Serotonergic Systems by Cannabinoids: Relevance to Cannabinoid-Induced Effects." Life Sci 192 (2018): 115–27. Print.

Metz, T. D., and E. H. Stickrath. "Marijuana Use in Pregnancy and Lactation: a Review of the Evidence." American Journal of Obstetrics and Gynecology (2015). Print.

Michalski, C. W., et al. "Cannabinoids in Pancreatic Cancer: Correlation with Survival and Pain." Int J Cancer 122.4 (2008): 742–50. Print.

Minerbi, A., W. Hauser, and M. A. Fitzcharles. "Medical Cannabis for Older Patients." Drugs Aging 36.1 (2019): 39–51. Print.

Minnesota Department of Health, Office of Medical Cannabis, "Review of Medical Cannabis Studies Relating to Chemical Compositions and Dosages for Qualifying Medical Conditions," 2018. Print.

Montane, E., et al. "Scientific Drug Information in Newspapers: Sensationalism and Low Quality. The Example of Therapeutic Use of Cannabinoids." Eur J Clin Pharmacol 61.5-6 (2005): 475–7. Print.

Morilak, D. A. "Modulating the Modulators: Interaction of Brain Norepinephrine and Cannabinoids in Stress." Exp Neurol 238.2 (2012): 145–8. Print.

Morrish, A. C., et al. "Protracted Cannabinoid Administration Elicits Antidepressant Behavioral Responses in Rats: Role of Gender and Noradrenergic Transmission." Physiol Behav 98.1-2 (2009): 118–24. Print.

Nasehi, M., et al. "Modulation of Cannabinoid Signaling by Amygdala Alpha2-Adrenergic System in Fear Conditioning." Behav Brain Res 300 (2016): 114–22. Print.

Neumeister, A., et al. "Elevated Brain Cannabinoid CB1 Receptor Availability in Post-Traumatic Stress Disorder: A Positron Emission Tomography Study." Mol Psychiatry 18.9 (2013): 1034–40. Print.

O'Connell, B. K., D. Gloss, and O. Devinsky. "Cannabinoids in Treatment-Resistant Epilepsy: A Review." Epilepsy Behav 70.Pt B (2017): 341–48. Print.

Orsini, A., et al. "Personalized Medicine in Epilepsy Patients." Journal of Translational Genetics and Genomics (2018). Print.

Pacher, P., and S. Steffens. "The Emerging Role of the Endocannabinoid System in Cardiovascular Disease." Semin Immunopathol 31.1 (2009): 63–77. Print.

Pamplona, F. A., L. R. da Silva, and A. C. Coan. "Potential Clinical Benefits of CBD-Rich Cannabis Extracts over Purified CBD in Treatment-Resistant Epilepsy: Observational Data Meta-Analysis." Front Neurol 9 (2018): 759. Print.

Paudel, K. S., et al. "Cannabidiol Bioavailability after Nasal and Transdermal Application: Effect of Permeation Enhancers." Drug Dev Ind Pharm 36.9 (2010): 1088–97. Print.

Pava, M. J., A. Makriyannis, and D. M. Lovinger. "Endocannabinoid Signaling Regulates Sleep Stability." PLoS One 11.3 (2016): e0152473. Print.

Pergolizzi Jr., J. A. Quang, J. F. Bisney. "Cannabinoid Hyperemesis." Medical Cannabis and Cannabinoids (2018). Print.

Pertwee, R. G. Handbook of Cannabis. Oxford, UK: Oxford University Press, 2014. Print.

Pham, Q. D., et al. "Chemical Penetration Enhancers in Stratum Corneum — Relation between Molecular Effects and Barrier Function." J Control Release 232 (2016): 175–87. Print.

Pisanti, S., and M. Bifulco. "Endocannabinoid System Modulation in Cancer Biology and Therapy." Pharmacol Res 60.2 (2009): 107–16. Print.

Poleszak, E., et al. "Cannabinoids in Depressive Disorders." Life Sci 213 (2018): 18–24. Print.

Pratt, M., et al. "Protocol for a Scoping Review of Systematic Reviews: Benefits and Harms of Medical Marijuana." Print.

Raikos, N., et al. "Determination of Delta9-Tetrahydrocannabinolic Acid A (Delta9-THCA-A) in Whole Blood and Plasma by LC-MS/MS and Application in Authentic Samples from Drivers Suspected of Driving under the Influence of Cannabis." Forensic Sci Int 243 (2014): 130–6. Print.

Rajesh, M., et al. "CB2-Receptor Stimulation Attenuates TNF-Alpha-Induced Human Endothelial Cell Activation, Transendothelial Migration of Monocytes, and Monocyte-Endothelial Adhesion." Am J Physiol Heart Circ Physiol 293.4 (2007): H2210-8. Print.

Richardson, K. A., A. K. Hester, and G. L. McLemore. "Prenatal Cannabis Exposure — the "First Hit" to the Endocannabinoid System." Neurotoxicol Teratol 58 (2016): 5–14. Print.

Romero-Sandoval, E. A., A. L. Kolano, and P. A. Alvarado-Vazquez. "Cannabis and Cannabinoids for Chronic Pain." Curr Rheumatol Rep 19.11 (2017): 67. Print.

Romigi, A., et al. "Cerebrospinal Fluid Levels of the Endocannabinoid Anandamide Are Reduced in Patients with Untreated Newly Diagnosed Temporal Lobe Epilepsy." Epilepsia 51.5 (2010): 768–72. Print.

Russo, E. B. "Cannabis and Epilepsy: An Ancient Treatment Returns to the Fore." Epilepsy Behav 70.Pt B (2017): 292–97. Print.

Russo, E. B. "Cannabis for Migraine Treatment: The Once and Future Prescription? An Historical and Scientific Review." Pain 76.1-2 (1998): 3–8. Print.

Russo, E. B. "Cannabis Therapeutics and the Future of Neurology." Front Integr Neurosci 12 (2018): 51. Print.

Russo, E. B. "Clinical Endocannabinoid Deficiency Reconsidered: Current Research Supports the Theory in Migraine, Fibromyalgia, Irritable Bowel, and Other Treatment-Resistant Syndromes." Cannabis Cannabinoid Res 1.1 (2016): 154–65. Print.

Russo, E. B. "Synthetic and Natural Cannabinoids: The Cardiovascular Risk." British Journal of Cardiology 22 (2015). Print.

Russo, E. B., G. W. Guy, and P. J. Robson. "Cannabis, Pain, and Sleep: Lessons from Therapeutic Clinical Trials of Sativex, a Cannabis-Based Medicine." Chem Biodivers 4.8 (2007): 1729–43. Print.

Sarid, N., et al. "Medical Cannabis Use by Hodgkin Lymphoma Patients: Experience of a Single Center." Acta Haematol 140.4 (2018): 194–202. Print.

Schier, A. R., et al. "Cannabidiol, a Cannabis Sativa Constituent, as an Anxiolytic Drug." Braz J Psychiatry 34 Suppl 1 (2012): S104–10. Print.

Scott, K. A., A. G. Dalgleish, and W. M. Liu. "The Combination of Cannabidiol and Delta9-Tetrahydrocannabinol Enhances the Anticancer Effects of Radiation in an Orthotopic Murine Glioma Model." Mol Cancer Ther 13.12 (2014): 2955–67. Print.

Sexton, M., C. Cuttler, and L. K. Mischley. "A Survey of Cannabis Acute Effects and Withdrawal Symptoms: Differential Responses across User Types and Age." J Altern Complement Med (2018). Print.

Shelef, A., et al. "Safety and Efficacy of Medical Cannabis Oil for Behavioral and Psychological Symptoms of Dementia: An Open-Label, Add-on, Pilot Study." J Alzheimers Dis 51.1 (2016): 15–9. Print.

Shen, J. J., et al. "Trends and Related Factors of Cannabis-Associated Emergency Department Visits in the United States: 2006–2014." J Addict Med (2018). Print.

Sherif, M., et al. "Human Laboratory Studies on Cannabinoids and Psychosis." Biol Psychiatry 79.7 (2016): 526–38. Print.

Sledzinski, P., et al. "The Current State and Future Perspectives of Cannabinoids in Cancer Biology." Cancer Med 7.3 (2018): 765–75. Print.

Slivicki, R. A., et al. "Positive Allosteric Modulation of Cannabinoid Receptor Type 1 Suppresses Pathological Pain without Producing Tolerance or Dependence." Biol Psychiatry 84.10 (2018): 722–33. Print.

Smith, P. A., et al. "Low Dose Combination of Morphine and Delta9-Tetrahydrocannabinol Circumvents Antinociceptive Tolerance and Apparent Desensitization of Receptors." Eur J Pharmacol 571.2–3 (2007): 129–37. Print.

Soares, V. P., and A. C. Campos. "Evidences for the Anti-Panic Actions of Cannabidiol." Curr Neuropharmacol 15.2 (2017): 291–99. Print.

Spierings, E. L. "Mechanism of Migraine and Action of Antimigraine Medications." Med Clin North Am 85.4 (2001): 943–58, vi–vii. Print.

Stampanoni Bassi, M., et al. "Exploiting the Multifaceted Effects of Cannabinoids on Mood to Boost Their Therapeutic Use against Anxiety and Depression." Front Mol Neurosci 11 (2018): 424. Print.

Stella, N. "Cannabinoid and Cannabinoid-Like Receptors in Microglia, Astrocytes, and Astrocytomas." Glia 58.9 (2010): 1017–30. Print.

Stinchcomb, A. L., et al. "Human Skin Permeation of Delta8-Tetrahydrocannabinol, Cannabidiol and Cannabinol." J Pharm Pharmacol 56.3 (2004): 291–7. Print.

Sulak, D. "Cannabis for Health Promotion and Disease Prevention." Medicinal Cannabis Conference 2017, Arcata Community Center, Arcata, California, April 29, 2017. Lecture.

Thomas, R. H., and M. O. Cunningham. "Cannabis and Epilepsy." Pract Neurol 18.6 (2018): 465–71. Print.

Tiwari, R. K., N. S. Chauhan, and H. S. Yogesh. "Ethosomes: A Potential Carries for Trans-dermal Drug Delivery." International Journal of Drug Development and Research 2.2 (2010). Print.

Torres, S., et al. "A Combined Preclinical Therapy of Cannabinoids and Temozolomide against Glioma." Mol Cancer Ther 10.1 (2011): 90–103. Print.

Touitou, E., et al. "Ethosomes — Novel Vesicular Carriers for Enhanced Delivery: Characterization and Skin Penetration Properties." J Control Release 65.3 (2000): 403–18. Print.

Touitou, E., et al. "Transdermal Delivery of Tetra-hydrocannabinol." International Journal of Pharmaceutics 43.1–2 (1988). Print.

Valiveti, S., et al. "In Vitro/in Vivo Correlation Studies for Transdermal Delta 8-THC Development." J Pharm Sci 93.5 (2004): 1154–64. Print.

Vandrey, R., et al. "Cannabinoid Dose and Label Accuracy in Edible Medical Cannabis Products." JAMA 313.24 (2015): 2491–3. Print.

Velasco, G., C. Sanchez, and M. Guzman. "Towards the Use of Cannabinoids as Antitumour Agents." Nat Rev Cancer 12.6 (2012): 436–44. Print.

Verbanck, P. "Short-Term and Long-Term Effects of Cannabis Use." Revue Medicale de Bruxelles 39.4 (2018): 246–49. Print.

Vigil, J. M., et al. "Effectiveness of Raw, Natural Medical Cannabis Flower for Treating Insomnia under Naturalistic Conditions." Medicines (Basel) 5.3 (2018). Print.

Vučković, S., et al. "Cannabinoids and Pain: New Insights from Old Molecules." Frontiers in Pharmacology (2018). Print.

Wade, D. T., et al. "A Preliminary Controlled Study to Determine Whether Whole-Plant Cannabis Extracts Can Improve Intractable Neurogenic Symptoms." Clin Rehabil 17.1 (2003): 21–9. Print.

Ware, M. A., et al. "Cannabis for the Management of Pain: Assessment of Safety Study (Compass)." J Pain 16.12 (2015): 1233–42. Print.

Weston-Green, K. "The United Chemicals of Cannabis: Beneficial Effects of Cannabis Phytochemicals on the Brain and Cognition." Recent Advances in Cannabinoid Research. Print.

Whiting, P. F., et al. "Cannabinoids for Medical Use: A Systematic Review and Meta-Analysis." JAMA 313.24 (2015): 2456–73. Print.

Whittle, B. A., G. W. Guy and P. Robson. "Prospects for New Cannabis-Based Prescription Medicines." Journal of Cannabis Therapeutics 1 (2001). Print.

Wilkinson, J. D., and E. M. Williamson. "Cannabinoids Inhibit Human Keratinocyte Proliferation through a Non-CB1/CB2 Mechanism and Have a Potential Therapeutic Value in the Treatment of Psoriasis." J Dermatol Sci 45.2 (2007): 87–92. Print.

Winkelman, M. "Archeological Investigations of Ancient Psychoactive Substances." Journal of Psychedelic Studies (2018). Print.

Woodhams, S. G., et al. "The Cannabinoid System and Pain." Neuropharmacology 124 (2017): 105–20. Print.

Yang, Y. T., and J. P. Szaflarski. "The US Food and Drug Administration's Authorization of the First Cannabis-Derived Pharmaceutical: Are We Out of the Haze?" JAMA Neurol (2018). Print.

Yesilyurt, O., et al. "Topical Cannabinoid Enhances Topical Morphine Antinociception." Pain 105.1-2 (2003): 303–8. Print.

Yin, A. Q., F. Wang, and X. Zhang. "Integrating Endocannabinoid Signaling in the Regulation of Anxiety and Depression." Acta Pharmacol Sin 40.3 (2019): 336–41. Print.

Zador, F., and M. Wollemann. "Receptome: Interactions between Three Pain-Related Receptors or the 'Triumvirate' of Cannabinoid, Opioid and Trpv1 Receptors." Pharmacol Res 102 (2015): 254–63. Print.

Zhou, D., et al. "Role of the Endocannabinoid System in the Formation and Development of Depression." Pharmazie 72.8 (2017): 435–39. Print.

Zias, J., et al. "Early Medical Use of Cannabis." Nature 363.6426 (1993): 215. Print.

Zimmerman, C. and E. Yarnell. "Herbal Medicines for Seizures." Alternative and Complementary Therapies 24.6 (2018). Print.

Zou, S., and U. Kumar. "Cannabinoid Receptors and the Endocannabinoid System: Signaling and Function in the Central Nervous System." Int J Mol Sci 19.3 (2018). Print.

RESOURCES

Ed Rosenthal's Marijuana Grower's Handbook by Ed Rosenthal

Feminist Weed Farmer: Growing Mindful Medicine in Your Own Backyard by Madrone Stewart

Handbook of Cannabis by Roger G. Pertwee, editor

Marijuana Garden Saver: A Handbook for Healthy Plants by J. C. Stitch, edited by Ed Rosenthal

Marijuana Horticulture Fundamentals: A Comprehensive Guide to Cannabis Cultivation and Hashish Production by K of Trichome Technologies

Marijuana Horticulture: The Indoor/Outdoor Medical Growers Bible by Jorge Cervan

Marijuana Pest and Disease Control: How to Protect Your Plants and Win Back Your Garden by Ed Rosenthal

Mycorrhizal Planet: How Symbiotic Fungi Work with Roots to Support Plant Health and Build Soil Fertility by Michael Phillips

The Cannabis Grow Bible: The Definitive Guide to Growing Marijuana for Recreational and Medical Use (2nd edition) by Greg Green

For testing of flowers and extracts: Nelson Analytical
www.nelsonanalytical.com

https://healer.com
Dr. Dustin Sulak

MEDICINE-MAKING TUTORIAL

Now that you understand how to work with cannabis, you can also access a free online video series to accompany the book at **https://tammisweet .com/book-bonus**. Get a peek behind the scenes while Tammi Sweet makes medicine, guides you through steps of the process, and addresses common areas where students have questions, concerns, and fears about making medicine with cannabis.

Get instant access to the free workshop at **https://tammisweet.com /book-bonus** for supplemental video material about:

- ◆ Which types of medicine to make and why

- ◆ How to make and use infused oils

- ◆ How to make and use tinctures

- ◆ Mistakes to avoid during the decarboxylation process

- ◆ How to decipher the Certificate of Analysis (COA)

- ◆ How to interpret lab results

- ◆ and of course . . . math. Because we all love math.

METRIC CONVERSIONS

Weight		
To convert	*to*	*multiply*
ounces	grams	ounces by 28.35
pounds	grams	pounds by 453.5
pounds	kilograms	pounds by 0.45

Volume		
To convert	*to*	*multiply*
teaspoons	milliliters	teaspoons by 4.93
tablespoons	milliliters	tablespoons by 14.79
fluid ounces	milliliters	fluid ounces by 29.57
cups	milliliters	cups by 236.59

INDEX

Page numbers in *italic* indicate images. Page numbers in **bold** indicate charts and tables.

CBN (cannabinol), 58–59, 131

cell membrane

  CB1 receptors at, 97, 100–101

  overview of, 91, *91*

central nervous system, 51, 77

chemical constituents, 48–49. *See also specific constituents*

chemotherapy, potentiation of, 220

chlorophyll, 64, 133

CHS. *See* cannabinoid hyperemesis syndrome

clinical applications

  anxiety, 199–201

  autism spectrum disorder, 182–184

  cancer, 218–225

  categories of, 165

  depression, 201–205, *203*

  dosage guidelines and, 165–166

  inflammation, chronic, 167–171, *168*, 172

  irritable bowel syndrome, 213–215

  ischemic brain injury, 181–182

  migraine, 215–217

  multiple sclerosis, 226–227

  nausea and vomiting, 210–213

  neurodegenerative diseases, 173–181

  neurological diseases, atypical, 182

  pain, 187–195, *188*

  posttraumatic stress disorder, 205–207

  schizophrenia, 207–208

  seizure disorders and epilepsy, 184–187

  sleep disorders, 208–210

cognition, contraindications and, 158

coir, 18

combustion, 128

commercial products, bioavailability and, 130

common nomenclature, 9

connective tissue, 79–80

contamination, 159

contraindications, 155–163

corticotropin-releasing hormone. *See* CRH

costs of pharmaceutical CBD, 187

COX-1 (cyclooxygenase 1), 170, *172*

COX-2 (cyclooxygenase 2), 171, *172*

CRH (corticotropin-releasing hormone), 75, *76*

criminalization, 11

cultivars

  chemicals and, 32

  defined, 28

  selecting, 123–124, 159, 211

  for sleep, 211

cyclooxygenase 1. *See* COX-1

cyclooxygenase 2. *See* COX-2

CYP450 (cytochrome P450) enzymes, 49

cytokines

  cannabidiol and, 57

  immune cells and, 228

  immune system and, 221–225

  inflammation and, 170–171, *172*

  inflammatory cascade and, 167–169

cytotoxic T cells (Tc), 224

D

dabs, 144–147

decarboxylation, 131, 134–137, *137*, 150

depletion, 159

depression, 159, 201–205, *203*

detoxification, 160

DHA (docosahexaenoic acid), 115

diarrhea, 160

dietary supplements, 25

digestive system, 80, 116

dimerization, 97

dioecious, 30

diseases (plant), 20–21

dispensaries, 123

F

FAAH (fatty acid amide hydrolase), 83
FABP (fatty acid binding protein), 83
Farm Bill (2018), 25
fascia, 80
fats, 116
fatty acid amide hydrolase. *See* FAAH
fatty acid binding protein. *See* FABP
FDA. *See* Food and Drug
    Administration
FECO (full-extract cannabis oil),
    144–147
female flowers, 30, *31*
fertilization, *44*
fight-or-flight response
    conditions of psyche and, 196–198,
        202-203, *203*
    overview of, 74–75, *76*
Figi, Charlotte, 184
first-pass metabolism, 53, 54, *54*
5-HT. *See* serotonin receptors
5-LOX (arachidonate 5-lipoxygenase),
    171, *172*
flavonoids. *See also specific flavonoids*
    medicine production and, 131
    overview of, 49, 63–64
flowers
    anatomy of, *31*
    banana, *40*
    life cycle of plant and, 45
    predicting sex of, *42*, 43
    seeded, 38–39, *39*
    unpollinated, *37*, 38
flushing, 19
Food and Drug Administration (FDA),
    25
freezing, 133
full agonists, 93
full-spectrum CBD oils, 12, 126
fungus, controlling, 20

G

GABA (gamma-aminobutyric acid),
    100, 106–107
Ganora, Lisa, 32
gardening, 15
gateway drugs, 15, 160–161
germination, 43
Gladstar, Rosemary, 13
glutamate, 198
glycerin, tincturing with, 139
glycine receptors, 107
GPR (G-protein coupled receptors),
    94, 100–101, 104
greenhouses, 18
growing, overview of, 16, 17–22

H

hashish, 28, 56
heat, 47, 134–136, *137*
helper T cells (Th), 224
hemp, labeling and, 25–26
hippocampus, 74–75, *76*, 196–198,
    202–203, *203*
history of cannabis medicine, 9–10
homeostasis (allostasis), 69
honey, cannabis-infused, 140
hops (*Humulus lupus*), 32
horticultural model, 16–17
horticulture, 17–22
humility, 15–16
humulene, 62
hydroponic growing, 18
hypervigilance, 74
hypotension, 161

I

IBD. *See* irritable bowel syndrome
ICAM. *See* intercellular adhesion
    molecule
immune cells, CB1 receptors at, 98

immune system
  cancer and, 221–225
  CB1 receptors and, 98
  CBD and, 57
  contraindications and, 161
  endocannabinoid system and, 70,
    77, 78–79, 221–225
  THC and, 51
indica, 26, 30. *See also Cannabis
  indica afghanica*
inflammation
  biochemistry of, 170–171, *172*
  CBD and, 57
  chronic vs. acute, 167
  inflammatory cascade and, 167, *168*
  mechanism of action against,
    168–169
  neurological, 173–174
  overview of, 166–167
  pain and, 193
  pharmaceutical model and, 169
infused oils, 141–143
inhalation, 52, 128–129
injuries, 77
insects, growth and, 20–21
insomnia, 162, 208–210
intake methods, 128–131
intercellular adhesion molecule
  (ICAM), 219
intermittent dosing, 120–121
intoxicants, defined, 55
inverse agonists, 93
irritable bowel syndrome (IBS),
  213–215
ischemic brain injury, 181–182
isolates, 12, 127

K

kaempferol, 64

L

labeling, 25–26, **27**, 126
leaves, 30, *31*
legalization, 11, 25
leukotrienes, 171, *172*
life cycle, overview of, 43–45
lighting, 21–22
limiting factors, 155–163
limonene, 62
linalool, 62
liver
  endocannabinoid system and, 80
  THC and, 49, 54, *54*
lock and key model, 92–93
5-LOX (arachidonate 5-lipoxygenase),
  171, *172*
lutein, 64
luteolin, 64
lysosomes, CB1 receptors at, 99

M

male plants
  flowers and, 30, *31*
  identifying, *42*, 43
  overview of, 30, 40–43, *41*
marijuana, 11
massage, 117
master plants, 6
math for medicine, 151–153
MCT (medium-chain trigylceride) oil,
  141–142
measurement, laws of, 152
Mechoulam, Raphael, 67
MED. *See* minimum effective dose
medicines. *See also* clinical
    applications
  contraindications for, 155–163
  decarboxylation and, 134-137, *137*
  dosage and, 147–153
  energetics of, 124
  infused oils and, 141–143

transient receptor potential cation
    channels. *See* TRP
trichomes
    chemistry of, 46, *47*
    decarboxylation and, 134–137
    flowering phase and, 45
    overview of, 30–31, 33–34, *34, 35, 37*
    terpenes and, 60–61
    whole-plant medicines and, 13–14
trigeminal vascular overexcitation, 216
trim, 32
TRP (transient receptor potential cat-
    ion channels), 111–113, *113*
TRPV (vanilloid) receptor agonists,
    190
TRPV (vanilloid) receptors, 112–113, *113*,
    114, 190
tumor growth, slowing of, 220
2AG (2-arachidonoylglycerol), 84, *84*

U
uric acid, 176

V
vanilloid (TRPV) receptors, 112–113, *113*,
    114, 190
vaporization, 128
vaporizers, 144

varieties. *See* cultivars
vegetative growth, *44*, 45
vitexin, 64
vomiting, 162, 210–213

W
wax, 144–147
well-being, 67–68, 70
Western medicine, 9–10
whole-plant extracts, 10, 126
whole-plant medicine, 7, 10, 13–14,
    120–121
wholistic approach, 14
withdrawal, 156, 200

X
xanthophyll, 64

Y
yields, 21–22

NOTES

NOTES

NOTES

NOTES

NOTES

NOTES